Advanced Design of Wastewater Treatment Plants:

Emerging Research and Opportunities

Athar Hussain
Ch. Brahm Prakash Government Engineering College, India

Ayushman Bhattacharya
Ch. Brahm Prakash Government Engineering College, India

A volume in the Advances in Environmental Engineering and Green Technologies (AEEGT) Book Series

IGI Global
DISSEMINATOR OF KNOWLEDGE

Published in the United States of America by
IGI Global
Engineering Science Reference (an imprint of IGI Global)
701 E. Chocolate Avenue
Hershey PA, USA 17033
Tel: 717-533-8845
Fax: 717-533-8661
E-mail: cust@igi-global.com
Web site: http://www.igi-global.com

Library of Congress Cataloging-in-Publication Data

Names: Hussain, Athar, 1976- author. | Bhattacharya, Ayushman, 1998- author.
Title: Advanced design of wastewater treatment plants : emerging research and opportunities / by Athar Hussain and Ayushman Bhattacharya.
Description: Hershey, PA : Engineering Science Reference, [2020] | Includes bibliographical references.
Identifiers: LCCN 2019001913| ISBN 9781522594413 (h/c) | ISBN 9781522594420 (eISBN) | ISBN 9781522594581 (s/c)
Subjects: LCSH: Water treatment plants--Design and construction.
Classification: LCC TD434 .H87 2020 | DDC 628.1/62--dc23 LC record available at https://lccn.loc.gov/2019001913

This book is published in the IGI Global book series Advances in Environmental Engineering and Green Technologies (AEEGT) (ISSN: 2326-9162; eISSN: 2326-9170)

British Cataloguing in Publication Data
A Cataloguing in Publication record for this book is available from the British Library.

All work contributed to this book is new, previously-unpublished material.
The views expressed in this book are those of the authors, but not necessarily of the publisher.

For electronic access to this publication, please contact: eresources@igi-global.com.

the amount of light penetrating the water, which reduces photosynthesis and the production of DO (S.K Garg, 2009).

Color

The color of the sewage indicates the freshness of sewage. If its color is greyish brown or yellowish, it indicates fresh sewage. With passage of time, as putrefaction starts it begins to get black. The color of stale and septic sewage is black (When all the oxygen has disappeared from sewage, it becomes septic). Other colors may also be formed due to presence of some specific industrial waste. The color of the sewage can normally be detected by the naked eye.

True color is the measurement made following the removal of colloidal or suspended sources of turbidity. The term "apparent color" includes not only color due to substances in solution, but also that due to suspended matter. Apparent color is determined on the original sample without filtration or centrifugation. In some highly colored industrial wastewaters color is contributed principally by colloidal or suspended material. In such cases both true color and apparent color should be determined.

Methods of Color Detection

Visual Comparison Method, color of the sample is determined by visual comparison with known concentration of colored solutions prepared by diluting stock platinum cobalt solution or properly calibrated glass colored disk is used for comparison. This method is useful for potable water and water in which color are due to naturally occurring materials, this method is not applicable to most highly colored industrial wastewater. Spectrophotometric method is applicable to potable and waste both domestic and industrial (Standard Method of Examination of Water and Wastewater, 1992). In this method light absorbed or transmitted is measured at dominant wavelength of a particular hue of sample. Spectrophotometer should have an effective operating range from 400 to 700 nm before measurement remove turbidity either by filtration or by centrifuging. Unit for color measurement is based on platinum cobalt scale.

Taste and Odor

Odor in domestic wastewater usually is caused by gases produced by the decomposition of organic matter or by substances added to the wastewater. Fresh, aerobic, domestic wastewater has been said to have the odor of kerosene or freshly turned earth, which is less objectionable than the odor of wastewater that has undergone anaerobic decomposition. Aged, septic sewage is considerably more offensive. The characteristics rotten-egg odor of hydrogen sulfide and the mercaptants is indicative of septic sewage, which is produced by the anaerobic microorganisms that reduce sulfate to sulfide. Hydrogen sulfide is a potent inhibitor of the cytochrome oxidase system, and thus inhibits oxidative phosphorylation, leading to cellular asphyxia (Dalefield, 2017). Industrial wastewater may contain either odorous compounds or compounds that produce odors during the process of wastewater treatment. The importance of odors at low concentrations in human terms is related to the psychological stress they produce rather than to the harm they do to the body. Offensive odors at higher concentration can cause poor appetite for food, lowered water consumption, impaired respiration, nausea and vomiting, and mental perturbation. One of the most common methods for measuring odor in water is the threshold odor test.

Temperature

Temperature of sewage is slightly more than that of water, because of the presence of industrial sewage. The temperature changes when sewage becomes septic because of chemical process. The lower temperature indicates the entrance of ground water into the sewage. Under general, conditions the temperature of the raw sewage is observed to be between 15 ℃ and 35 ℃ at various places in different seasons. Optimum temperatures for bacterial activity are in the range from 25℃ to 35℃. Aerobic digestion and nitrification stops when the temperature rises to 500℃. When temperature drops to about 15 ℃, methane-producing bacteria become quite inactive, and at about 5℃, the autotrophic-nitrifying bacteria practically stop functioning. In general, biological treatment activity accelerates in warm temperatures and slows in cool temperatures, but extreme hot or cold can stop treatment processes altogether. Therefore, some systems are less effective during cold weather and some may not be appropriate for very cold climates. Wastewater temperature also affects receiving waters. Hot water, for example, which is a byproduct of many manufacturing processes, can be a pollutant. When discharged in large quantities, it can raise the temperature of receiving streams locally and disrupt the natural balance of aquatic life.

Chemical Characteristics of Wastewater

Organic Substances

Carbohydrates

Carbohydrates, which are widely distributed in nature and found in wastewater, are organic substances that include starch, cellulose, sugars, and wood fibers; they contain carbon, hydrogen, and oxygen. Sugars are soluble but starches are insoluble in water. In lower organisms (e.g., bacteria), carbohydrates are utilized to synthesize fats and proteins as well as a source of energy. In the absence of oxygen, the end products of decomposition of carbohydrates are organic acids and alcohols, as well as gases such as carbon dioxide and hydrogen sulfide. Of the carbohydrates, glucose, sucrose and lactose are the major ones with the smaller proportions of galactose, fructose, xylose and arabinose. Together they account for 90-95% of all the carbohydrate present which is equivalent to 50-120mg/L. A diurnal variation in carbohydrate concentration and composition is evident, and although glucose accounts for over 50% of the total carbohydrate content in composite samples, sucrose concentration is greater than glucose in the afternoon. The ration of hexose to pentose is between 10 and 12. The non soluble high molecular weight carbohydrates such as starches, cellulose and wood fibre are restricted to the suspended solids fraction resulting in a low hexose to pentose ratio (2.0-2.6) and a concentration of 30-38 mg/L (Nick, 2004).

Proteins

Proteinaceous materials constitute a large part of the wastewater biosolids; biosolids particles that do not consist of pure protein will be covered with a layer of protein that will govern their chemical and physical behavior. Moreover, the protein content ranges from 15 to 30% of the organic matter present for digested biosolids and 28 to 50% in the case of activated biosolids. Proteins and urea are the chief sources of nitrogen in wastewater. When proteins are present in large quantities, microorganisms decompose and produce end products that have objectionable foul odors. During this decomposition process, proteins are hydrolyzed to amino acids and then further degraded to ammonia, hydrogen sulfide, and organic compounds.

Fats, Oils and Waxes

Fats, oils, waxes, and other related constituents found in wastewater are commonly grouped under the category of grease. Fats and oils contributed to domestic wastewater include butter, lard, margarine, and vegetable fats and oils. Fats, which are compounds of alcohol and glycerol, are among the more stable of organic compounds and are not easily decomposed by bacteria; however, they can be broken down by mineral acids, resulting in the formation of fatty acid and glycerin. Normal concentration ranges for fats in domestic wastewater are between 40-100 mg/L. Fats are broken down by hydrolytic action to yield fatty acids and a wide variety of free fatty acid is reported from sewage, including all the saturated ones from C_8 to C_{14} as well as C_{16}, C_{18}, and C_{20}. The major acids include palmitic, stearic and oleic which form between 75-90% of the present (Painter, 1971).

Measurement of Organic Matter

Biochemical Oxygen Demand (BOD)

Biochemical Oxygen Demand (BOD) is a chemical procedure for determining how fast biological organisms use up oxygen in a body of water. It is used in water quality management and assessment, ecology and environmental science. In analytical chemistry, quantitative analysis is the determination of the absolute or relative abundance (often expressed as a concentration) of one, several or all particular substance(s) present in a sample. BOD is not an accurate quantitative test, although it is considered as an indication of the quality of a water source (Abdulla et al, 2012). It is most commonly expressed in milligrams of oxygen consumed per litre of sample during 5 days of incubation at 20°C or 3 days of incubation at 27°C. The BOD test must be inhibited to prevent oxidation of ammonia. If the inhibitor is not added, the BOD will be between 10% and 40% higher than can be accounted for by carbonaceous oxidation.

Stages of Decomposition in the BOD Test

1. There are two stages of decomposition in the BOD test: a carbonaceous stage and a nitrogenous stage.
   a. The carbonaceous stage represents oxygen demand involved in the conversion of organic carbon to carbon dioxide.
   b. The second stage or the nitrogenous stage represents a combined carbonaceous plus nitrogenous demand, when organic nitrogen, ammonia and nitrite are converted to nitrate. Nitrogenous oxygen demand generally begins after about 6 days.Under some conditions, if ammonia, nitrite, and nitrifying bacteria are present, nitrification can occur in less than 5 days. In this case, a chemical compound that prevents nitrification is added to the sample if the intent is to measure only the carbonaceous demand. The results are reported as carbonaceous BOD (CBOD) or as $CBOD_5$ when a nitrification inhibitor is used.

BOD Dilution Method

BOD is the amount of oxygen (Dissolved Oxygen (DO)) required for the biological decomposition of organic matter. The oxygen consumed is related to the amount of biodegradable organics. When organic substances are broken down in water, oxygen is consumed

$$\text{Organic Carbon} + O_2 \rightarrow CO_2 \quad (1)$$

where, organic carbon in human waste includes protein, carbohydrates, fats, etc. Measure of BOD = Initial oxygen- Final Oxygen after (5 days at 20°C) or (3 days at 27°C).

Two standard 300 mL BOD bottles are filled completely with wastewater. The bottles are sealed. Oxygen content (DO) of one bottle is determined immediately. The other bottle is incubated at 20°C for 5 days or (or at 27°C for 3 days) in total darkness to prevent algal growth. After which its oxygen content is again measured. The difference between the two DO values is the amount of oxygen consumed by micro-organisms during 5 days and is reported as BOD_5.

Since the saturated value of DO for water at 20°C is 9.1 mg/L only and that the oxygen demand for wastewater may be of the order of several hundred mg/L, therefore, wastewater are generally diluted so that the final DO in BOD test is always ≥ 2 mg/L. Precaution is also taken so as to obtain at least 2 mg/L change in DO between initial and final values.

$$BOD_5 = \frac{(DO_i - DO_f)}{P} \tag{2}$$

where, DO_i and DO_f are initial and final DO concentrations of the diluted sample, respectively. P is called as dilution factor and it is the ratio of sample volume (volume of wastewater) to total volume (wastewater plus dilution water). In the above formula, it was assumed that the diluted wastewater had no oxygen demand of itself and that the dilution wastewater used was pure.

Most of the times, microorganisms are added in the dilution water (seeded water) so as to have enough microorganisms for carrying out biodegradation of organic waste. In this case, the oxygen demand of seeded water is subtracted from the demand of mixed sample of waste and dilution water.

For seeded samples

$$BOD_5\left(in\, mg / L\right) = D^* \left[\left[\underbrace{(DO)_{t=0} - (DO)_{t=5}}_{Sample}\right] - \left[\underbrace{(DO)_{t=0} - (DO)_{t=5}}_{Blank<0.2mg/L, may\, be\, neglected}\right]\right] \tag{3}$$

f = ratio of seed volume in dilution solution to seed volume in BOD test on seed, normally f is near to 1

where D* = dilution factor

For unseeded samples the difference between the two DO values is the amount of oxygen that is consumed by micro-organisms during the 5 days and is reported as the BOD_5 (5day BOD) value of the sample.

$$BOD_5\left(in\, mg / L\right) = D^* \left[\left[\underbrace{(DO)_{t=0} - (DO)_{t=5}}_{Sample}\right] - \left[\underbrace{(DO)_{t=0} - (DO)_{t=5}}_{Blank}\right]\right] \tag{4}$$

where D* = dilution factor

If the waste is strong, the oxygen in the bottle will run out before the end of the incubation period. This is why we dilute the waste. If we over-dilute it, the change of DO will be too small to be statistically reliable. A typical graph of DO in the BOD bottle as a function of time might look like Figure 1.

As shown in the graph, with mounting time, the bacteria consume all of the organic material, and the DO stops dropping. The total DO drop at this point represents the "ultimate BOD" (also called BOD_u or sometimes BOD_L for "limiting BOD"). Twenty to thirty days is a long time to wait, so usually the bottles are opened and measured after 5 days, and the test results are reported as the "5-day BOD" or $BOD_{5.}$ Figure 2 represents the relationship between measured BOD versus time.

Derivation of BOD First Order Reaction

Assuming that the rate of decomposition of organic waste is proportional to the waste remaining at any time t is governed by a first- order function as given below:

$$\frac{dL_t}{dt} = -kL_t \tag{5}$$

where, L_t is the amount of oxygen demand left after time t and k is the BOD rate constant ($time^{-1}$). Solving this equation yields

Figure 1. Relationship between dissolved oxygen and time
(S.K Garg, 2009)

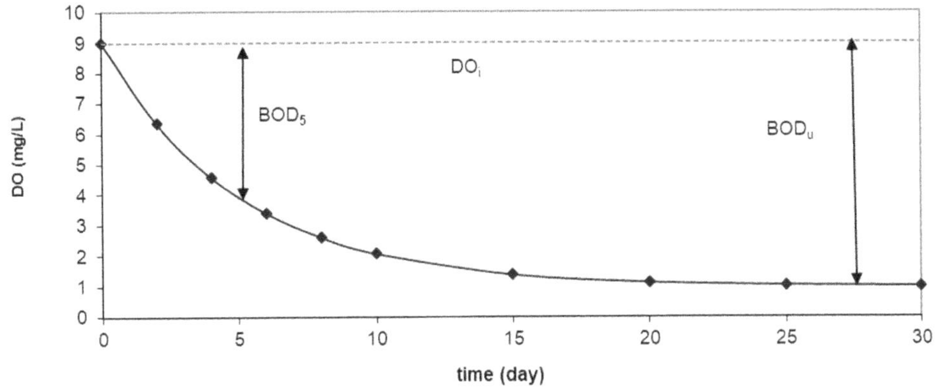

Figure 2. Relationship between biological oxygen demand and time (Sperling, 2007)

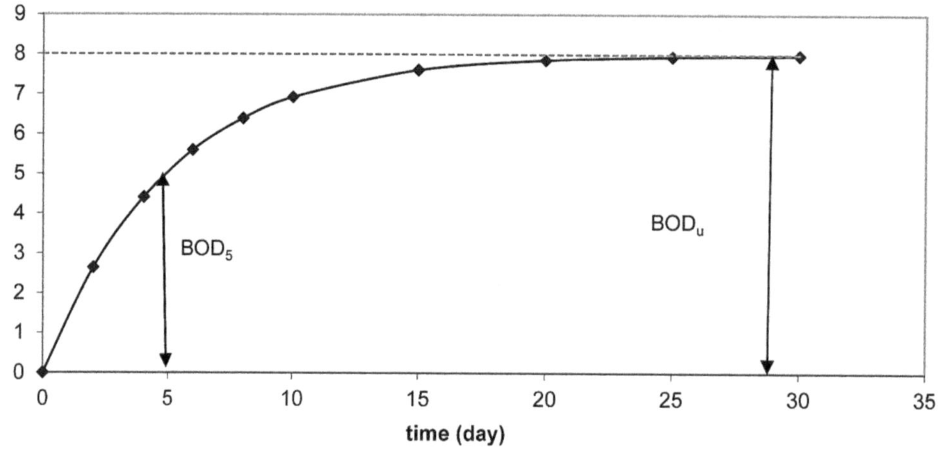

$$\int \frac{dL_t}{L_t} = \int -K.dt \tag{6}$$

$$\log_e L_t = -K.t + C \tag{7}$$

when t=0 i.e at start L_i=L. Substituting in above equation we get,

$$\log_e L = K(0) + C \tag{8}$$

$$C = \log_e L \tag{9}$$

Substituting value of C in equation (g), we get,

$$\log_e L_t = -K.t + \log_e L \tag{10}$$

$$\log_e \frac{L_t}{L} = -K.t \tag{11}$$

$$2.3\log_{10}\frac{L_t}{L} = -K.t \tag{12}$$

$$\log_{10}\frac{L_t}{L} = -\frac{K.t}{2.3} = -0.434K.t \tag{13}$$

Using 0.434 K=K_D, K_D is the Deoxygenation constant or BOD rate constant

$$\log_{10}\frac{L_t}{L} = -K_D.t \tag{14}$$

$$\frac{L_t}{L} = 10^{-KD.t} \tag{15}$$

Now, L is the ultimate carbonaceous oxygen demand and it is also the amount of O_2 demand left initially (at time 0, no DO demand has been exerted, so BOD = 0) At any time, $L_o = BOD_t + L_t$ (that is the amount of DO demand used up and the amount of DO that could be used up eventually). If Y_t represents the total amount of organic matter oxidised in t days (i.e the BOD of t days), then we have

$$Y_t = L - L_t \tag{16}$$

$$Y_t = L\,[1 - \frac{L_t}{L}] \tag{17}$$

$$\frac{L_t}{L} = 1 - \frac{Y_t}{L} \tag{18}$$

Substituting value of $\frac{L_t}{L}$ from eq(15)

$$1 - \frac{Y_t}{L} = 10^{-KD.t} \tag{19}$$

The overall equation for BOD calculation is

$$Y_t = L\left[1-(10)^{-K_D.t}\right] \tag{20}$$

where Y_t = total amount of organic matter oxidised in time 't'

K_D = De-oxygenation constant its value ranges for different types of wastewaters (d^{-1})

The value of K_D however determines the speed of the BOD reaction without influencing the ultimate BOD. It is found to vary with temperature of sewage and this relationship is given by equation:

$$\mathbf{K_D} = \mathbf{K_{D(20^\circ)}}(1.047)^{T-20^\circ}$$

$K_{(D20^\circ)}$= Deoxygenation constant at 20°C, its numerical values varies from 0.05 to 0.2 per day depending upon the nature of organic matter . simple compounds such as sugars and starches are easily utilized by micro-organisms and have a high K_D rate while complex molecules such as phenols are difficult to assimilate and hence have allow K_D values. Some values are tabulated in Table 1. $K_{(T^\circ)}$ = Deoxygenation constant at temperature T°C.

Table 1. Typical value of K_D at 20°C for various types of waters and wastewaters

S. No.	Water Type	K_D Value per Day
1	Tap Waters	< 0.05
2	Surface waters	0.05 – 0.1
3	Municipal Waters	0.1 – 0.15
4	Treated sewage Effluents	0.05 – 0.1

Chemical Oxygen Demand (COD)

This test is carried out on the sewage to determine the extent of readily oxidizable organic matter, which is of two types:

1. Organic matter which can be biologically oxidized is called biologically active.
2. Organic matter which cannot be oxidized biologically is called biologically inactive.
    a. COD is a measure of the oxygen equivalent of the organic matter content of a sample that is susceptible to oxidation by a strong chemical oxidant.
    b. It is an indirect measurement of the amount of organic compounds in wastewater, and is often approximately 2-½ times the BOD.
    c. It is expressed in milligrams per liter (mg/L), which indicates the mass of oxygen consumed per liter of solution.

Analytical Procedure

$$\text{Organic C} + Cr_2O_7\text{-}\ CO_2 + H_2O + Cr_2O_4^{2-} \tag{21}$$

A sample is refluxed in strongly acidic solution with a known excess of potassium dichromate ($K_2Cr_2O_7$) for 2-3 h. After digestion, the remaining unreduced $K_2Cr_2O_7$ is titrated with ferrous ammonium sulphate to determine the amount of $K_2Cr_2O_7$ consumed. Then, the oxidizable matter is calculated in terms of oxygen equivalent. This procedure is applicable to COD values between 40 and 400 mg/L.

Differences Between BOD and COD

COD always oxidize things that the BOD cannot or will not measure; therefore, COD is always higher than the BOD. The common compounds which cause COD to be higher than BOD include sulfides, sulfites, thiosulfates and chlorides. The general relationship between BOD and COD for sewage and most human wastes is about 1 unit of BOD≈0.64–0.68 units of COD. The relationship is not consistent and it may vary considerably for industrial wastewaters.

Theoretical COD

Theoretical oxygen demand (ThOD) is the calculated amount of oxygen required to oxidize a compound to its final oxidation products.

CxHyOz+ ¼•(4x+y-2z)•O2 → x•CO2 + y/2•H2O (22)

So,

COD = 8(4x+y-2z)/(12x+y+16z) g COD/g

BOD/COD Ratio

COD of waste water will always be more than its BOD_u value, since there difference will represent the quantum of non biodegradable organics (NBOs) present in wastewater. The BOD_u/COD ratio will therefore always be less than 1.0; but this value shall approach towards 1.0 with decreasing amount of NBOs. If this ratio is found to be between 0.92-1.0, the wastewater can be considered to be virtually fully biodegradable. Since BOD_u is generally not measured and BOD_5 is measured and only BOD_5 is measured the ratio BOD_5/COD, usually referred to as BOD/COD ratio becomes important. Since BOD_5 is about 68% of BOD_u we can easily say that BOD_5/COD ratio should for fully biodegradable wastewaters vary between 0.92 x 0.68 = 0.63 to 1.0 x 0.68 = 0.68.

INORGANIC SUBSTANCES

Chlorides

Concentration of chlorides in sewage is greater than the normal chloride content of water supply.

Sources of chlorides in natural waters include: (1) leaching of chloride from rocks and soils; (2) in coastal areas, saltwater intrusion; (3) agricultural, industrial, domestic, and human wastewater; and (4) infiltration of groundwater into sewers adjacent to saltwater. The normal chloride content of sewage is

120 mg/L, whereas the permissible limit of chloride content in water is 250 mg /L.

Nitrogen

The presence of nitrogen in sewage is an indication of the presence of the organic matter and may occur in one or more of the following forms: Free ammonia called ammonia nitrogen, Albuminoid or Organic Nitrogen, Nitrites, Nitrates. The free ammonia indicates the very first stage of decomposition of organic matter (thus indicating recent pollution); albuminoid nitrogen indicates the quantity of nitrogen in sewage before the decomposition of organic matter. Nitrates, indicates the presence of fully oxidized organic matter in sewage. The nitrites, thus indicates the intermediate stage of conversion of organic matter of sewage into stable forms, thus indicating the progress of treatment. Their presence shows that the treatment given to the sewage is incomplete, and sewage is stale. Whereas, the presence of nitrates indicates the well oxidized and treated sewage. Generally nitrogen content in the untreated sewage is observed to be in the range of 20 to 50 mg/L measured as TKN. The principal nitrogen compounds in domestic sewage are proteins, amines, amino acids, and urea.

Phosphorus

The main sources of phosphorus released into the environment include fertilizers, detergents, cleaning preparations and boiler waters to which phosphates are added for treatment. It exists in three forms: organic phosphorus (associated with organic molecules), orthophosphate- (exists as an anion) and polyphosphates (from detergents). Only Orthophosphate can be chemically precipitated, however, most of the organic phosphorus and polyphosphates are converted to the orthophosphate form during biological treatment (Krishnaswamy, 2009). The concentration of PO_4 in raw sewage is generally observed in the range of 5 to 10 mg/L. Phosphorus, however, is required in extremely small concentrations to sustain algae growth. Undesirable algal growth has been reported where inorganic phosphorus levels were in the range of 0.01 to 0.05 mg/L.

Sulphides, Sulphates and Hydrogen Gas

Sulfide-containing waste streams are generated by a number of industries such as petrochemical plants, tanneries, viscose rayon manufactures, the gasification of coal for electricity production, or by the anaerobic treatment of sulfate containing wastewaters. It is emitted into the environment as dissolved sulfide (S^{2-} and HS^{-}) in wastewaters and as H_2S in waste gases. Hydrogen sulfide is generated in relatively stagnant wastewater systems as a result of the biological breakdown of sulfates (SO_4^{2-}) in anaerobic wastewater environments as shown in the following biochemical reaction (Qaisar, 2014).

$$SO_4^{2-} + \text{Organic Matter} + \text{Anaerobic Bacteria} \rightarrow S^{2-} + H_2O + CO_2 \quad (23)$$

Sulfate is present in great abundance in municipal wastewater systems and primarily stems from household cleaning detergents. Once the anaerobic bacteria reduce the sulfates to sulfide (S^{2-}), it reacts with hydrogen to produce hydrogen sulfide according to the following reaction:

$$S^{2-} + 2H^{+} \rightarrow H_2S \quad (24)$$

Hydrogen sulfide prefers to exist in the gas phase; thus, once produced in the aqueous phase, it will rapidly partition to the gas phase once the headspace in the pipe is available. Once hydrogen sulfide is present in the headspace of the pipe, aerobic bacteria lining the interior surface of the pipeline convert the hydrogen sulfide into sulfuric acid as shown in the following biochemical reaction.

$$H_2S + 2O_2 + \text{Aerobic Bacteria} \rightarrow H_2SO_4 \quad (25)$$

pH

pH is a measure of hydrogen ion concentration in water. In other words, it is the acid or alkaline condition of water. Several factors like temperature, aeration and input from external source also interfere with the pH. Several studies have found that the pH of the wastewater had an important effect on turbidity removal efficiency (Mandal, 2014). The highest removal efficiency was found at higher pH. Even without the addition of coagulant a considerable part of the turbidity would precipitate at elevated pH. If the pH value is less than 7, the sewage is acidic and if the pH vale is more than 7, the sewage is alkaline. The fresh sewage is alkaline, with passage of time pH tends to fall due to production of acid by bacterial action in anaerobic or nitrification processes. However, with treatment of sewage, the pH tends to rise.

Solids

Solid materials in wastewater can consist of organic and/or inorganic materials and organisms. The solids must be significantly reduced by treatment or they can increase BOD when discharged to receiving waters. The typical values of solids found in wastewater is depicted in Table 2.

Total Solids (TS)

Total solids include both the suspended solids and the dissolved solids which are obtained by separating the solid and liquid phase by evaporation. Total solids (TS) in wastewater is the amount of all solids, which are determined by drying a known volume of the sample in a preweighed crucible dish at 105 °C. After cooling in an exsiccator, the crucible dish is again weighed. TS is determined by using the following formula:

$$TS = \frac{(M_1 - M_2)}{V} \tag{26}$$

with

Figure 3. Variation of pH values in wastewater

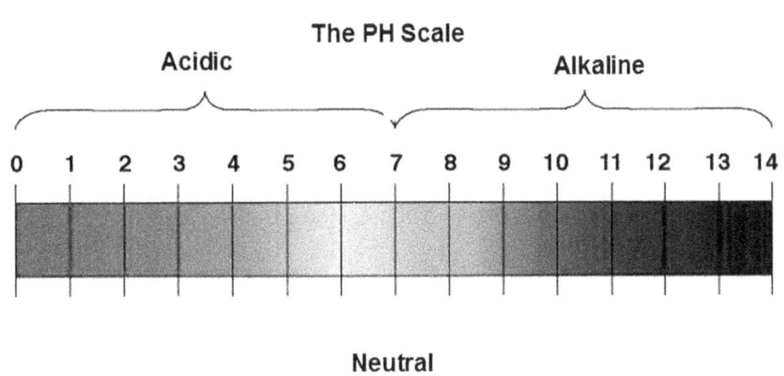

M_1: Mass of crucible dish after drying at 105 °C (mg)
M_2: Mass of initial crucible dish (mg)
V: Volume of sample (L)

Total Suspended Solid (TSS)

Total Suspended Solids are the solids retaining in a filter and is usually determined by filtration using glass fibre filters. In all analytical procedures for determination of suspended solids, weighed filters are used for sample filtration, the filters are dried at about 105°C after filtration, cooled in an desiccator to room temperature and the weight of the loaded filter is determined. SS is determined by using the following formula:

$$SS = \frac{\left(M_4 - M_5\right)}{V} \tag{27}$$

with

M_4: Mass of filter after drying at 105 °C (mg)
M_5: Mass of initial filter (mg)
V: Volume of sample (L)

Volatile Suspended Solids (VSS)

Solids that can be volatilized and burned off when TSS are ignited at 500 ± 50°C. The volatile solids represent an estimate of organic matter in the solids, while the non volatile solids (fixed) represent the inorganic or mineral matter

$$VSS = \left(\frac{M_4 - M_6}{V}\right) \tag{28}$$

M_4: Mass of filter after drying at 105 °C (mg)
M_6: Mass of filter after ignition at 550 °C (mg)
V: Volume of sample (L)

Total Dissolved Solids (TDS)

Can be determined by subtracting suspended solids from total solids. The solids passing through the filter consist of colloidal and dissolved solids.

Settleable Solids

Solid settle at the bottom of an "Imhoff Cone" after 60mins and expressed as mL/L. The fraction that does not settle represents the non settleable solids.

Total Organic Carbon

The total carbon analyzer allows a total soluble carbon analysis to be made directly on an aqueous sample. In many cases TOC can be correlated with COD and occasionally with BOD values. As the time required for carbon analysis is generally short, such correlations are extremely helpful when monitoring treatment plant flows for efficiency control.

Biological Characteristics of Wastewater

The principal groups of microorganisms found in wastewater are bacteria, fungi, protozoa, microscopic plants and animals, and viruses. Most microorganisms (bacteria, protozoa) are responsible and are beneficial for biological treatment

Figure 4. Characterization of solids

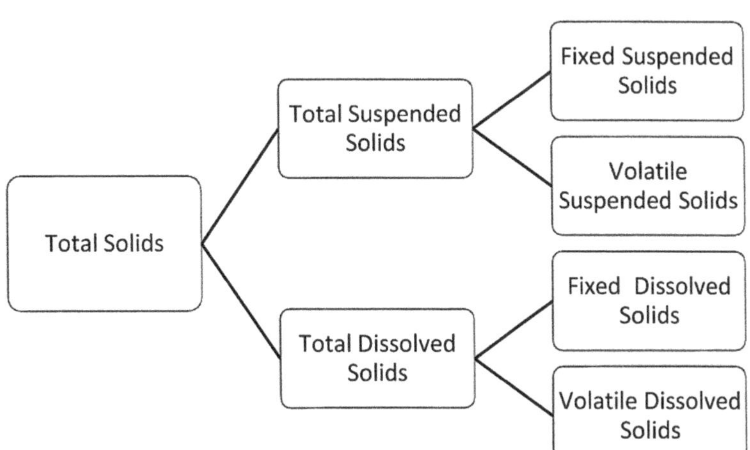

Table 2. General standards for discharge of environmental pollutants: effluents

Parameter (mg/L)	Inland Surface Water	Public Sewer	Land for Irrigation
Suspended Solids, Max	100	600	200
Oil and Grease Max.	10	20	10
pH Values	5.5-9.0	5.5-9.0	5.5-9.0
Oil and Grease mg/L	10	20	10
Total Kjeldahl Nitrogen, (TKN) mg/L	100	-	-
Free Ammonia(NH_3)	5.0	-	-
Total Residual Chlorine Max.	1	-	-
BOD	30	350	100
COD	250	-	-
Mercury (Hg)	0.01	0.01	-
Lead (Pb)	0.1	1	-
Ammoniacal Nitrogen (as N)	50	50	-
Cadmium (Cd)	2	1	-
Copper (Cu)	3	3	-
Zinc (Zn)	5	15	-

(Source: Environment Protection Rules, 1986)

processes of wastewater. Pathogenic organisms are usually excreted by humans from the gastrointestinal tract and discharge to wastewater. Water-borne disease includes cholera, typhoid, paratyphoid fever, and diarrhoea. The number of pathogenic organisms in wastewaters is generally low in density and they are difficult to isolate and identify. Therefore, indicator bacteria such as total coliform (TC) and fecal coliform (FC) are used as indicator organisms.In order to determine whether water has been contaminated by fecal material, a series of tests are used to demonstrate the presence or absence of coliforms.

The coliform group is comprised of Gram-negative, nonspore-forming, aerobic to facultative anaerobic rods, which ferment lactose to acid and gas. Two organisms in this group include E. coli and Enterobacter aerogenes; however, the only true fecal coliform is E. coli, which is found only in fecal material from warm-blooded animals. The presence of this organism in a water supply is evidence of recent fecal contamination and is sufficient to order the water supply closed until tests no longer detect E. coli. There are two methods for determining the presence and density of coliform bacteria. The membrane filter (MF) technique provides a direct count of colonies trapped and then cultured. The multiple-tube fermentation technique is a three-stage procedure in which the results are statistically expressed in terms of the Most Probable Number (MPN). These stages - the presumptive stage, confirmed stage, and completed test are briefly summarized below.

- **Presumptive Stage:** A series of lauryl tryptose broth primary fermentation tubes are inoculated with graduated quantities of the sample to be tested. The inoculated tubes are incubated at 35 ± 0.5°C for 24 + 2 hr, at which time the tubes are examined for gas formation. For the tubes in which no gas is formed, continue incubation and examine for gas formation at the end of 48 ± 3 hour. Formation of gas in any amount within 48 ± 3 hr is a positive presumptive test. Confirmed Stage: The confirmed stage is used on all primary fermentation tubes showing gas formation during the 24-hr and 48-hr periods. Fermentation tubes containing brilliant green lactose bile broth are inoculated with medium from the tubes showing a positive result in the presumptive test. Inoculation should be performed as soon as possible after gas formation occurs. The inoculated tubes are incubated for 48 ± 3 hr at 35 ± 0.5°C. Formation of gas at any time in the tube indicates a positive confirmed test.
- **Completed Test:** The completed test is performed on all samples showing a positive result in the confirmed test. It can also be used as a quality control measure on 20% of all samples analyzed. One or more plates of eosin methylene blue are streaked with sample to be analyzed. The streaked plates are incubated for 24 ± 2 hr at 35 ± 0.5°C. After incubation, transfer one or more typical colonies (nucleated, with or without metallic sheen) to a lauryl tryptose broth fermentation tube and a nutrient agar slant. The fermentation tubes and agar slants are incubated at 35 ± 0.5°C for 24 ± 2 hr, or for 48 ± 3 hr if gas is not produced. From the agar slants corresponding to the fermentation tubes in which gas formation occurs, gram-stained samples are examined microscopically. The formation of gas in the fermentation tube and the presence of gram-negative, non-spore-forming, rod-shaped bacteria in the agar culture may be considered a satisfactorily completed test, demonstrating the positive presence of coliform bacteria in the analyzed sample.

NUMERICAL PROBLEMS

Examples 1

The BOD_5 of wastewater is 170mg/L at 20°C. The k value is known to be 0.23per day. What would BOD_8 be, if the test was run at 15°?

Sol. $BOD_{5,}$ $Y_5 = 170$mg/L

$k = 0.23$

$K_D = 0.434k = 0.434 \times 0.23$

$= 0.0998 \cong 0.1$

$$Y_t = L\left[1-(10)^{-K_D.t}\right]$$

$$Y_5 = L\left[1-(10)^{-K_D.5}\right]$$

$$170 = L\left[1-(10)^{-K_D.5}\right]$$

$$170 = L\left[1-(10)^{-0.1\times5}\right]$$

$$L = \frac{170}{0.683} = 248.90 \text{ mg/L}$$

Now, we have to find K_D at 15°C.

$$K_{D(T^O)} = K_{D(20^\circ)}(1.047)^{T-20^\circ}$$

$$K_{D(15^O)} = 0.1(1.047)^{15-20^\circ}$$

$= 0.1(1.047)^{-5}$

$= 0.079$

Now again using,

$Y_t = L\left[1-(10)^{-K_D.t}\right]$, where Y_t is BOD of t days

$$Y_8 = 248.90\left[1-(10)^{-0.079\times 8}\right]$$

$$Y_8 = 248.90\left[1-(10)^{-0.632}\right]$$

$= 248.90 \times 0.766$

$BOD_8 = 190.65 mg/L$

Examples 2

The BOD of a sewage incubated for one day at 30°C has been found to be 130 mg/L. What will be the 5-day 20°C BOD? Assume K_1=0.1 at 20°C.

Sol. $K_{D(T^O)} = K_{D(20^\circ)}(1.047)^{T-20^\circ}$

$K_{D(30^O)} = 0.1(1.047)^{30-20^\circ}$

$= 0.158$

$$Y_t = L\left[1-(10)^{-K_D.t}\right]$$

$130 = L\left[1-(10)^{-0.158\times 1}\right]$ [For one day, t=1]

$$L = \frac{130}{0.304} = 427.63 \text{ mg/L}$$

Therefore,

$$Y_t = L\left[1-(10)^{-K_D.t}\right],$$

$$\text{BOD at } 20^0\text{C}, Y_5 = 427.63\left[1-(10)^{-0.1\times5}\right]$$

$= 427.63\times [1- 0.316]$

$= 292.49\text{mg/L}$

Example 3

COD analysis of a waste water sample is carried out through chemical method using (N/10) K-dichromate solution as reducing agent. Excess dichromate is titrated against ferrous ammonium sulphate. 100 ml of waste water sample is used. The titter value of ferrous ammonium sulphate are 55 ml and 80 ml for original sample and blank sample respectively. Determine the COD of the sample. Assume there is no interfering element in the sample.

Sol. COD = 8000(b-s) × n/sample volume

where b is the volume of FAS used in the blank sample, s is the volume of FAS in the original sample, and n is the normality of FAS.

COD = 8000(80-55) × 0.1/100 = 200 mg/L

Examples 4

Determine the 5-day BOD for a 15 ml sample that is diluted with dilution water to a total volume of 300 ml when the initial DO concentration is 10 mg/L and after 5 days, has been reduced to 2 mg/L.Sol. $D_0 = 10$

$D_5 = 2$

$P = 15$ ml/300ml $= 0.05$

$$\text{BOD (mg/L)} = \frac{D_o - D_5}{P} = 160\text{mg/L}$$

Examples 5

Organic carbon, $C(H_2O)$, and ammonia nitrogen, NH_3-N, are oxidized to carbon dioxide (CO_2) and nitrate (NO_3), respectively, by the bacteria which are naturally present in wastewater and in natural systems such as lakes and rivers. Both of these reactions consume oxygen and thus cause a negative impact on water quality. The amount of oxygen theoretically required to oxidize a carbonaceous or nitrogenous waste (its ThOD) may be calculated according to the stoichiometry of the reactions as outlined in class. The production of coke, a fuel produced from coal for use in steel mills, generates a waste stream rich in ammonia, phenol, and naphthalene. Calculate the carbonaceous ThOD and the nitrogenous ThOD of a waste containing 25 mg/L of ammonia nitrogen (NH_3-N), 50 mg/L of phenol (C_6H_5OH) and 150 mg/L of naphthalene ($C_{10}H_8$).

Sol. The theoretical nitrogenous oxygen demand, ThNOD, is estimated from the stoichiometry of the nitrification reaction,

$$NH_3 + 2O_2 \rightarrow NO_3^- + H_2O + H^+ + \Delta$$

where it can be calculated that 4.57 mgO_2 are required for each $mgNH_3$-N oxidized. Thus,

Box 1.

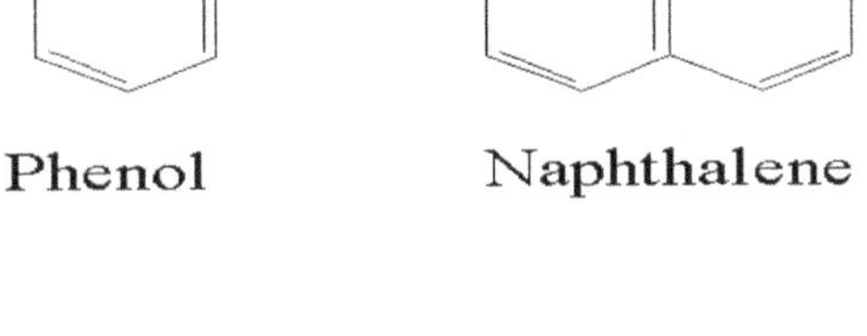

$$ThNOD = 4.57 \cdot \frac{mgO_2}{mgNH_3 - N} \cdot 25 \frac{mgNH_3 - N}{L} = 114 \frac{mgO_2}{L}$$

For the carbonaceous oxygen demand, ThCOD, we need to balance the equations. First for phenol,

$$C_6H_5OH + 7O_2 \rightarrow 6CO_2 + 3H_20 + \Delta$$

Above equation ensue that 7 moles or 224 gO_2 is required for each mole of phenol. Therefore,

$$ThCOD = \frac{50mg\,phenol}{L} \cdot \frac{mmole\,phenol}{94mg\,phenol} \cdot \frac{224mgO_2}{mmole\,phenol} = 119 \frac{mgO_2}{L}$$

and for napthalene,

$$C_{10}H_8 + 12O_2 \rightarrow 10CO_2 + 4H_2O + \Delta$$

We require 12 moles or 384 gO_2 for each mole of napthalene. Therefore,

$$ThCOD = \frac{150mg\,napthalene}{L} \cdot \frac{mmole\,napthalene}{128mg\,napthalene} \cdot \frac{384mgO_2}{mmole\,napthalene} = 450 \frac{mgO_2}{L}$$

The overall ThCOD is 450 + 119 = 569 $\frac{mgO_2}{L}$ and the total theoretical oxygen demand (ThOD) of the waste is 114 + 569 = 683 $\frac{mgO_2}{L}$.

REFERENCES

Abdulla, H.J., Al-Quraeshi, N.K., & Al-Awadi, F.N. (2012). Study of Chemical Oxygen Demand (COD) in Relation to Biochemical Oxygen Demand (BOD). *Journal of Kerbala University, 10*(3).

Dalefield, R. (2017). Smoke and Other Inhaled Toxicants. In *Veterinary Toxicology for Australia and New Zealand*. Elsevier. doi:10.1016/B978-0-12-420227-6.00019-0

Garg, S.K. (2009). Sewage Disposal and Air Pollution Engineering. *Environmental Engineering*, (2).

Gray Nick, F. (2004). Biology of wastewater Treatment (2nd ed.). Academic Press. doi:10.1142/p266

Henze & Comeau. (2008). *Wastewater Characterization, Biological Wastewater Treatment: Principles Modelling and Design*. IWA Publishing. Retrieved from https://ocw.un-ihe.org/pluginfile.php/462/mod_resource/content/1/Urban_Drainage_and_Sewerage/5_Wet_Weather_and_Dry_Weather_Flow_Characterisation/DWF_characterization/Notes/Wastewater%20characterization.pdf

Inamori, Y., & Fujimoto, N. (n.d.). *Point Source of Pollution, Water Quality and Standards –Vol II, Encyclopedia of Life Support Systems (EOLSS)*. Retrieved from https://www.eolss.net/sample-chapters/c07/E2-19-04-01.pdf

Krishnaswamy, U., Muthusamy, M., & Perumalsamy, L. (2009). Studies on the Efficiency of the Removal of Phosphate Using Bacterial Consortium for the Biotreatment of Phosphate Wastewater. *European Journal of Applied Sciences, 1*(1).

Mandal, H. K. (2014). Influence of Wastewater PH on Turbidity. *International Journal of Environmental Research and Development*, *4*(2), 105–114.

Muttamara, S. (1996). Wastewater characteristics, Resources. *Conservation and Recycling*, *16*(1-4), 145–159. doi:10.1016/0921-3449(95)00052-6

Odlare, M. (2014). *Introduction. In Water Resources, Reference Module in Earth Systems and Environmental Science*. Elseiver. doi:10.1016/B978-0-12-409548-9.09035-7

Painter, H.A. (1971). *Water and water pollution handbook*. Academic Press.

Qaisar, M., Ping, Z., Jing, C., Yousaf, H., Jaffar, H. M., Dong-lei, W., & Bao-lan, H. (2007). Sources of sulfide in waste streams and current biotechnologies for its removal. *Journal of Zhejiang University. Science A*, *8*(7), 1126–1140. doi:10.1631/jzus.2007.A1126

Russell, D. L. (2006). *Practical wastewater treatment*. Hoboken, NJ: John Wiley & Sons, Inc. doi:10.1002/0470067926

Spellman, F. R. (2013). *Handbook of water and wastewater treatment plant operations*. Taylor and Francis Group. doi:10.1201/b15579

Sperling, M. V. (2007). *Biological Wastewater Treatment* (Vol. 1). IWA publishing.

Standard Method of Examination of Water and Wastewater. (1992). *American Public Health Association,American Water Works Association*. Water Environment Federation.

Section 2

Reactor Analysis

Chapter 2
Reactor Analysis

ABSTRACT

Reactor is a device or vessel within which chemical processes are carried out for experimental or manufacturing purposes. The most common basic types of reactors are tanks (where the reactants mix in the whole volume) and pipes or tubes (for laminar flow reactors and plug flow reactors). Both types can be used as continuous reactors or batch reactors, and either may accommodate one or more solids (reagents, catalysts, or inert materials), but the reagents and products are typically fluids (liquids or gases). Reactors in continuous processes are typically run at steady-state, whereas reactors in batch processes are necessarily operated in a transient state. When a reactor is brought into operation, either for the first time or after a shutdown, it is in a transient state, and key process variables change with time. The purpose of this chapter is to discuss the types of the reactors used in the wastewater treatment, modeling ideal, non-ideal flows in the reactor and treatment kinetics. Furthermore, the chapter considers both kinetic and hydrodynamic aspect while designing the reactor.

TYPES OF REACTORS USED IN THE WASTEWATER TREATMENT

- Batch reactor
- Continuous stirred-tank reactor (CSTR)
- Plug flow reactor (PFR)

DOI: 10.4018/978-1-5225-9441-3.ch002

Batch Reactor

The batch reactor is the generic term for a type of vessel widely used in the process industries. Its name is something of a misnomer since vessels of this type are used for a variety of process operations such as solids dissolution, product mixing, chemical reactions, batch distillation, crystallization, liquid/liquid extraction and polymerization.

Batch reactors are normally used for small-scale operation, testing new processes, the manufacture of expensive products, and processes difficult to convert to continuous. The advantage is that high conversions can be achieved due to leaving the reactants in reactor. The disadvantages are high labour costs, variability of products (batch to batch), and they are difficult to operate/automate for large-scale production.In order to model a batch reactor, we need to make the assumption that at any given time the reactor is well-mixed so that the composition, temperature, and pressure are the same everywhere in the reactor.

Figure 1. Schematic diagram of batch reactor

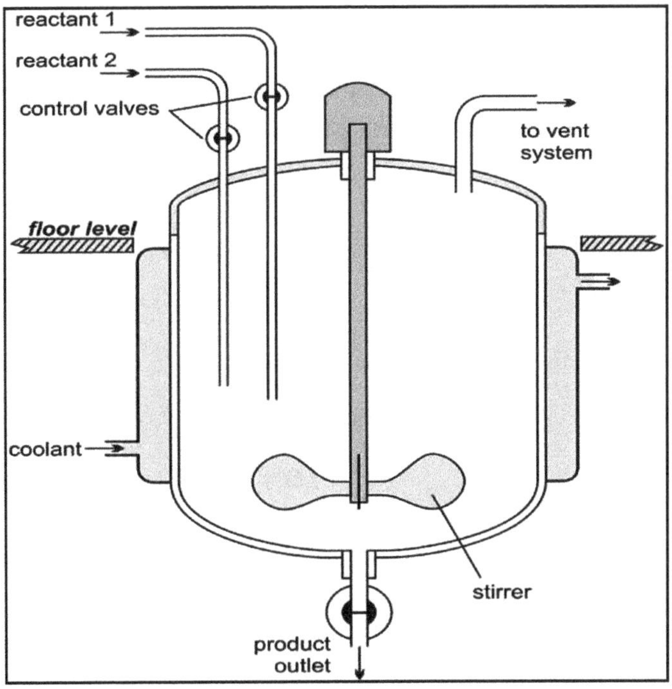

When the reaction takes place in a liquid (or sometimes solids), very often the reacting component occurs at a much lower concentration then the principal component of the liquid (i.e. the solvent). In this case, a good approximation is that the volume of the liquid and pressure in the reactor remain constant with time as there is only a small change in the density of the liquid during the reaction. For gas-phase reactions, the entire volume of the reactor is filled by the gas, so that the reaction volume is equal to the reactor volume (whereas for a liquid, the reaction volume is the volume of the liquid which is less than the reactor volume). Thus for a gas-phase reaction, if the reaction either consumes or generates moles, the net effect will be a change to the pressure in the reactor, because the volume remains constant.

If the reaction generates moles, the pressure in the reactor will increase, whereas if moles are consumed, the pressure in the reactor will decrease. Because we assume that the batch reactor is well mixed (remember the mole balance must be made over a volume element which is spatially uniform with respect to composition and temperature) we can apply the mole balance over the entire volume of the reactor (Thomas Rodgers, 2013) . There is no inflow or outflow in a batch Reactor (Krishna, 2013).

Complete Mix Reactor

In completely mixed reactors, the content within the control volume remains in completely mixed (homogeneous) state. Such system often needs a stirring device to ensure the mixing in the reactors and are commonly known as Continuous Stirred Tank Reactor (CSTR). This makes the temperature, concentration, and reaction rate independent of position in the reactor

Figure 2. Schematic diagram of complete mix reactor

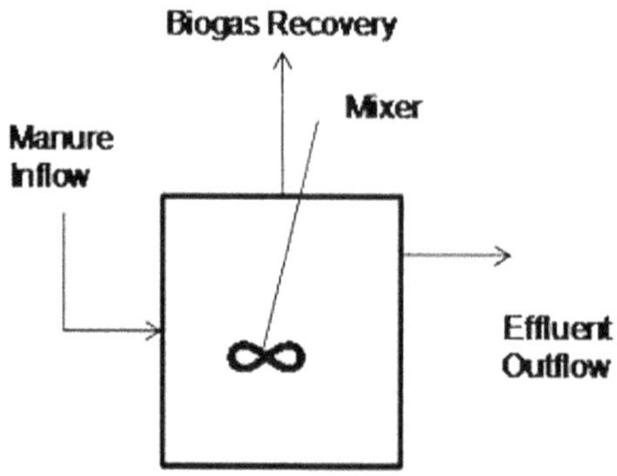

(Rosen, 2014). In continuous flow reactors, there is a continuous inflow to and outflow from control volume. Therefore, the rate of mass flux in and out has to be considered in mass balance. CSTRs are simply well -mixed tanks which are used to model well-mixed environmental reservoirs. In CSTRs, the concentration of a substance in outflow remains equal to that in the reactor.

Plug Flow Reactor

Fluid particles pass through the reactor and are discharged in the same sequence in which they entered the reactor. Each fluid particle remains in the reactor for a time period equal to the theoretical detention time. This type of flow is approximated in long tanks with a high length/width ratio in which longitudinal dispersion is minimal or absent. In an ideal plug flow reactor, it is assumed that there is no mixing of the medium along the long axis (X-axis) of the reactor although there may be lateral mixing in the medium at any point along the long axis (ie the Y-axis). Plug flow reactors are used for some of the following applications:

- Large-scale production
- Fast reactions
- Homogeneous or heterogeneous reactions
- Continuous production
- High-temperature reactions

Figure 3. Schematic diagram of plug flow reactor
(Harrou et al, 2017)

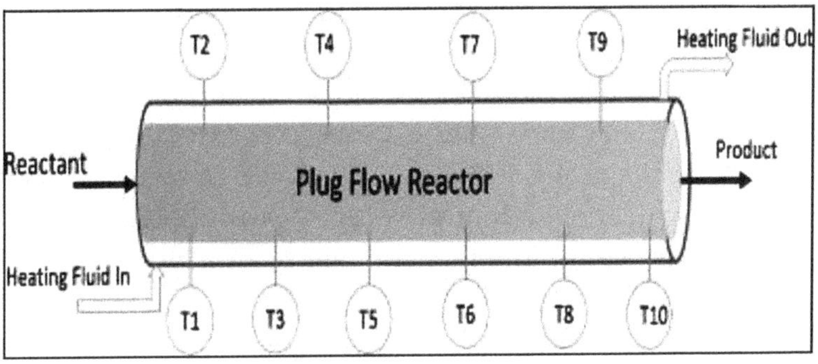

Mass Balance Analysis

Mass Balance is an application of conservation of mass: mass can neither be produced nor destroyed. The accounting of all mass in a process/system confined under a control volume is referred as mass (or material) balance. The mathematical way of describing the mass balance is with mass conservation equations which state/assume that "what goes into the system must either come out of the system somewhere else, get used up or generated by the system, or remain in the system and accumulate" (Stenstrom and Rosso, 2013).

Mass accumulation = MassInput – Mass output + Mass generation – Massconsumption

Principle of Conservation of Mass

Mass accumulation rate = Mass flux in – Mass flux out ± Net rate of chemical production

$$\frac{dm}{dt} = m_f(\text{in}) - m_f(\text{out}) \pm \frac{dm}{dt}_{\text{reaction}} \quad (1)$$

Concept of Mass Balance: Control Volume

A mass balance is only meaningful in terms of a specific region of space usually called the Control Volume, which has boundaries across which the terms mass flux in and mass flux out could be determined. For the transformation reactions, the mass of contaminants present in the control volume, at any given time, is considered as the amount of that substance available for reaction. In wastewater treatment systems, usually, the volume of specific reactor units (tank sizes) are considered as control volume. With-in the control volume, the flow of the mass could follow a completely mixed flow model or the plug-flow model

MODELING IDEAL FLOW IN REACTORS

Mass Balance for CSTR

Conservative System, Pulse Input (Slug Dose)

See Box 1.

Rate of change of tracer in a fixed volume = Inflow-Outflow

$$\frac{\mathrm{d}(\mathrm{CV})}{\mathrm{dt}} = \mathrm{QC_o} - \mathrm{QC}$$

$$\frac{\mathrm{d}(\mathrm{C})}{\mathrm{dt}} = - \frac{Q}{V}\,\mathrm{C}$$

$$\int \frac{d(C)}{C} = - \frac{1}{\theta}\int dt \; [\text{Detention time}, \theta = \frac{Volume, V}{Discharge, Q}]$$

$$\ln \mathrm{C} = - \frac{t}{\theta} + \mathrm{A} \qquad (2)$$

At $t=0$, $C=C_o$

$A = \ln C_o$

therefore in eq.2

Box 1.

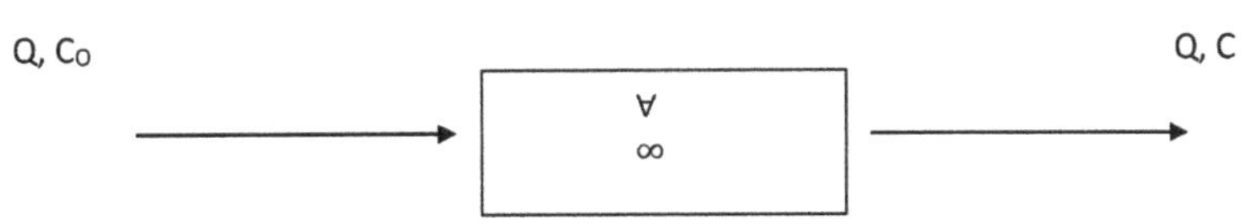

$$\ln C - \ln C_o = -\frac{t}{\theta}$$

$$\ln \frac{C}{C_0} = -\frac{t}{\theta}$$

$$\mathbf{C} = \mathbf{C_o}\, e^{-\left(\frac{t}{\theta}\right)} \qquad (3)$$

Conservative System, Step Input

$$\frac{d(CV)}{dt}. = QC_O - QC$$

$$\frac{d(C)}{dt} = \frac{Q}{V}(C_o\text{-}C)$$

$$\frac{d(C)}{dt} = \frac{1}{\theta}(C_o\text{-}C)$$

$$\int \frac{d(C)}{C_0 - C} = -\frac{1}{\theta}\int dt$$

$$-\ln (C_o\text{-}C) = \frac{t}{\theta} + A \qquad (4)$$

At t=0,C=0

A=-ln C_0

Therefore in eq. 15

$$-\ln (C_o - C) = \frac{t}{\theta} - \ln C_o$$

$$\frac{C - C_0}{C_0} = e^{\frac{-t}{\theta}}$$

$$\mathbf{C = C_0(1 - e^{\frac{-t}{\theta}})} \tag{5}$$

Non Conservative, Step Input

Rate of change of concentration = Inflow – Outflow – Degradation

$$\frac{d(CV)}{dt} = QC_o - QC - KCV$$

$$\frac{dC}{dt} = \frac{QC_o - QC - KCV}{V}$$

$$\frac{dC}{dt} + KC = \frac{C_o - C}{\theta}$$

$$\frac{dC}{dt} = \frac{C_o - C}{\theta} + KC$$

$$\frac{dC}{dt} = \frac{C_o}{\theta} - C(\frac{1}{\theta} + k)$$

$$\int \frac{dC}{\frac{C_o}{\theta} - C\frac{1 + K\theta}{\theta}} = \int dt$$

$$\frac{\ln(\frac{C_o}{C} - C\left(\frac{1+K\theta}{\theta}\right)}{\frac{1+K\theta}{\theta}} = \text{t} + \text{A}$$

At t=0, C=0

Therefore

$$\text{A} = \frac{\ln(\frac{\text{C}_o}{\theta})}{-\left(\frac{1+K\theta}{\theta}\right)}$$

$$\frac{\ln(\frac{\text{C}_o}{\theta} - C\left(\frac{1+K\theta}{\theta}\right)}{-\left(\frac{1+K\theta}{\theta}\right)} = \text{t} + \frac{\ln(\frac{C_o}{\theta})}{-\left(\frac{1+k\theta}{\theta}\right)}$$

$$\mathbf{C} = \frac{C_O}{K\theta+1}\left(1-e^{1-t\frac{(K\theta+1)}{\theta}}\right) \tag{6}$$

Steady State (No Accumulation w.r.t Time)

$$\frac{d(CV)}{dt} = \text{QC}_o - \text{QC} - \text{KCV}$$

$$0 = \text{QC}_o - \text{QC} - \text{KCV}$$

$$\text{Q}(\text{C}_o - \text{C}) = \text{KCV}$$

$$\text{C} = \frac{C_{0-C}}{\theta K}$$

$$\mathbf{C} = \frac{C_o}{1 + K\theta} \tag{7}$$

for “n” number of CSTR are connected in series then

$$\mathbf{C} = \frac{C_o}{(1 + K\theta)^n} \tag{8}$$

Mass Balance for PFTR

See Box 2.

Mass balance for the plug shown above: V. dC/dt= QC(x) –QC(x+dx) ±V. $dC/dt|_{reaction}$

In Steady State [dC/dt= 0]:

$$QC(x+dx) - QC(x) = \pm V. \frac{dC}{dt}|_{reaction}$$

$$Q[C(x+dx) - C(x)] = \pm A. dx. \frac{dC}{dt}|_{reaction}$$

$$\frac{\left[C\left(x + dx\right) - C\left(x\right)\right]}{dx} = \pm \left(\frac{A}{Q}\right). \frac{dC}{dt}|_{reaction}$$

Box 2.

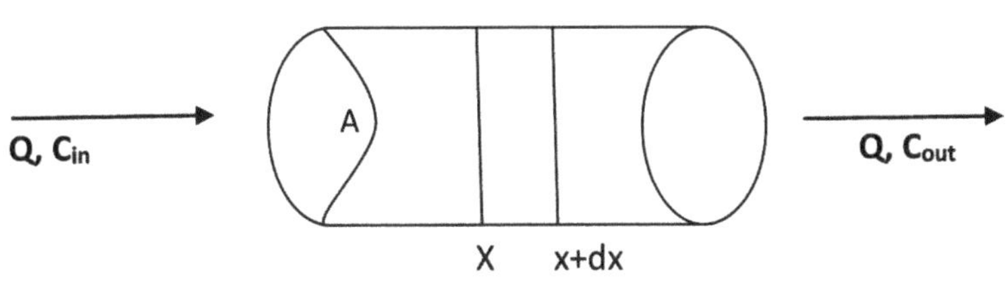

$$\frac{dC}{dx} = \pm \left(\frac{A}{Q}\right) . \frac{dC}{dt}\Big|_{\text{reaction}}$$

>>Steady State; Ist Order decay at rate constant k:

$$\frac{dC}{dx} = - \left(\frac{A}{Q}\right) . \text{KC}$$

$$\frac{dC}{C} = - K\left(\frac{A}{Q}\right) . \text{dx}$$

$$\text{C} = \text{C}_{\text{in}} . \exp\left(- \frac{KV}{Q}\right) = \text{C}_{\text{in}} . \exp(-\text{kt}) \tag{9}$$

Reactions, Reaction Rates and Reaction Rate Coefficients

The reaction rate represents how fast a chemical component is converted into another by a chemical reaction. More specifically, the reaction rate, r, is the moles of substance formed (or appearing) per unit volume per unit time (mol m^{-3} s^{-1}). The reaction rate is negative when the moles of substance are being consumed by the reaction (as occurs for reactants before any product has been formed). The reaction rate is positive when the moles of substance increase with time due to the reaction (as occurs for products before any product has been formed)

A reaction equation is an algebraic equation that is solely a function of the properties of the reacting materials and reaction conditions (e.g. species concentration, temperature, pressure, or type of catalyst) at any point in the system. (Box 3)

Box 3.

$$\text{A} \longrightarrow \text{Products}$$

The reaction rate may be a linear function of concentration, i.e. $-r_A = kC_A$ or may be some other algebraic function of concentration, such as $-r_A = kC^2_A$ or $-r_A = (k_1C_A) / (1 + k_2C_A)$.

For a given reaction, the concentration dependence of the rate must be determined from experimental observation. Typical examples of reaction rates are a first order reaction given by, $-r_A = kC_A$ or a second order reaction, $-r_A = kC^2_A$ where k is called the rate constant, which varies with temperature. The order of the reaction with respect to a component j corresponds to the exponent of concentration j in the rate law. Usually the order of the reaction provides some insight into the molecular mechanism for the reaction. A first order reaction corresponds to a uni-molecular process, whereas a second order reaction corresponds to reaction controlled by collisions between molecules. These rules are only strictly true when the reaction is an elementary step. Most reactions are combinations of elementary steps, which can lead to more complicated rate laws. In addition, the rate law can depend on the relative concentrations of the components. For instance, rate laws are independent of the concentrations of components, which occur at a large excess relative to another component (i.e. water).

NUMERICALS

Q-1 A pollutant is following first order decay in the lake water and its half–life was determined to be 120 days. Calculate the rate constant and the time required to achieve 10% of its initial concentration. Let's consider the lake as batch system as no inflow and outflow information is provided.

Sol. For first Order decay at rate constant k: $\frac{dC}{dx} = -kC$; $C = C_0e^{-kt}$

Half life is 120 days, i.e. C remains $\frac{C_0}{2}$ in t =120 days.

$$\frac{C_0}{2} = C_0e^{-k(120)}$$

$$\frac{1}{2} = e^{-k(120)}$$

$$\ln\left(\frac{1}{2}\right) = -k \qquad (120)$$

$$k = \frac{-\ln(0.5)}{120} = \frac{-(-0.693)}{120} = \frac{0.693}{120}$$

Rate constant, $k = 0.00577\ d^{-1}$

Time required for achieving 10% of its initial concentration ($C = 0.1C_0$) is say t.

$$0.1C_0 = C_0 e^{-0.00577(t)}$$

$$t = -\ \frac{\ln(0.1)}{0.00577} = 399 \text{ days}$$

Q-2 A 400 m3 CSTR receives water from a single inlet at 40 m3/h flow containing 20 mg/L of total Polycyclic Aromatic Hydrocarbon (TPAHs). Determine the steady state TPAHs concentration in the single exit stream, if TPAHs are degraded at first–order kinetics with

$k = 0.27\ h^{-1}$. What should be the volume of a corresponding plug flow channel if the same degree of pollutant reduction is needed (with unchanged flow rate and 'k')?

Sol. For a Steady State, CSTR ($C_{out} = C$) with first order decay $\frac{dC}{dt}|reaction = -\ kC$;

$$0 = Q \times C_{in} - Q \times C - V.kC$$

$$(Q + kV)\ C = Q \times C_{in}$$

$$C = C_{in}\left(\frac{Q}{Q + kV}\right)$$

$$C = C_{in}\left(\frac{1}{1 + \frac{kV}{Q}}\right)$$

The Steady State TPH conc. C = 20mg/L. {1/(1+0.27 h^{-1}×400 m^3/40 m^3h^{-1})}

$= 20\text{mg/L}. (\frac{1}{3}\times 7) = 5.4\ \text{mg/L}$

Q-3 Find maximum flow rate of wastewater that can be maintained in a aerated lagoon of volume 1 million gallon to treat the wastewater (having BOD of 450 mg/l) at 25°C so that the BOD of treated stream does not exceed 90 mg/l. The reaction coefficient k is 0.308 d^{-1} at 25°C, and the value of Ø is 1.06.

Sol. Given: S = 90 mg/l; S_0=450mg/l; V=1 M gallon

$$\frac{S}{S_0} = \frac{1}{1 + K\theta}$$

$$\frac{90}{450} = \frac{1}{1 + 0.308\times\frac{1}{Q}}$$

Q = 0.077 M gallon/day.

BIOFILM PROCESS

Moving Bed Biofilm Reactor

Moving bed biofilm reactor (MBBR) is a type of wastewater treatment process that was first invented by Prof. Hallvard Odegaard at Norwegian University of Science and Technology in the late 1980s. The MBBR system consists of an aeration tank (similar to an activated sludge tank) with special plastic carriers that provide a surface where a biofilm can grow. The biofilm grows on small high density polyethylene media which are in constant motion in the reactor. The biofilm carriers are made of high density polyethylene (0.95-0.98g/cm3) and usually shaped like small cylinders with a cross inside and longitudinal fins outside (Odegaard, 1999). An example is high-density polyethylene (HDPE) which has a density close to 0.95 g/cm^3. The carriers will be mixed in the tank by the aeration system and thus will have good contact between the substrate in the influent wastewater and the biomass on the carriers.

To prevent the plastic carriers from escaping the aeration it is necessary to have a sieve on the outlet of the tank.

Figure 4. Schematic diagram of moving bed biofilm reactor

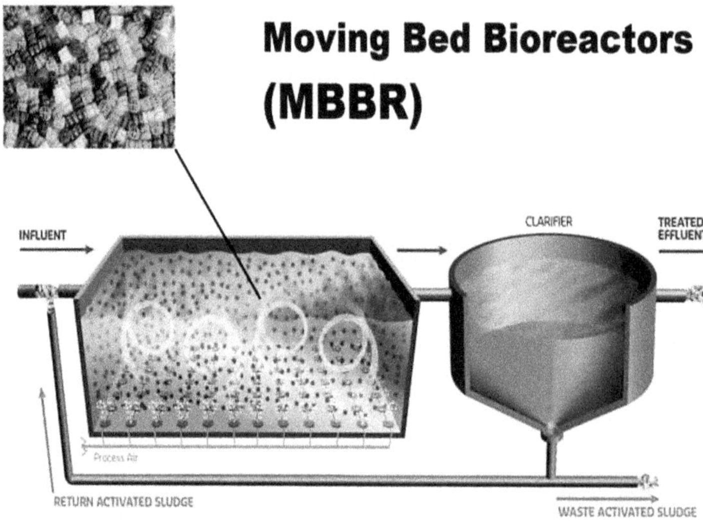

Figure 5. Schematic diagram of plastic media

MBBR ADVANTAGES

(1)It consists of compact units with small size, thus accommodating less area, (2) High treatment capacity in terms of wastewater treatment, (3) Solids removal efficiency is high, (4) Settling characteristics is enhanced, (5) Operation at higher suspended biomass (6) Concentrations resulting in long sludge retention times (7) Enhanced process stability, (8) Low head loss, (9) No filter channeling, (10) There is no need of periodic backwashing, (11) Reduced sludge production and no problems with sludge bulking

Anaerobic Process

In anaerobic systems, most of the biodegradable organic matter present in the waste is converted to biogas (about 70-90%), which is removed from the liquid phase and leaves the reactor in a gaseous form. Only a small portion of the organic material is converted into the microbial biomass (about 5-15%), which then constitutes the excess sludge in the system.

Anaerobic Baffled Reactor

An Anaerobic Baffled Reactor (ABR) is an improved septic tank because of the series of baffles over which the incoming wastewater is forced to flow. The increased contact time with the active biomass (sludge) results in improved treatment. The majority of settleable solids are removed in the sedimentation chamber at the beginning of the ABR, which typically represents 50% of the total volume. The up-flow chambers provide additional removal and digestion of organic matter: BOD may be reduced by up to 90%, which is far superior to that of a conventional septic tank. As sludge is accumulating, desludging is required every 2 to 3 years. Critical design parameters include a hydraulic retention time (HRT) between 48 to 72 hours, up-flow velocity of the wastewater less than 0.6m/h and the number of up-flow chambers (2 to 3).

Design Principles

1. ABRs start with settling chamber for larger solids and impurities followed by series of at least two, sometimes up to five up-flow chambers
2. Hydraulic Retention Time (HRT) is relatively short and varies from only a few hours upto two or three days.

Figure 6. Schematic Diagram of Anaerobic Baffled Reactor

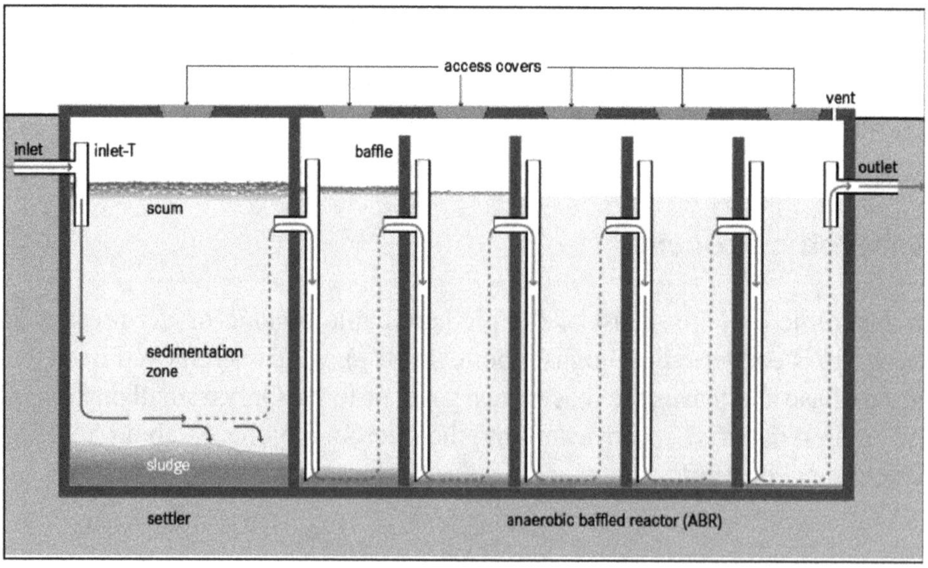

3. Up-flow velocity is the most crucial parameter for dimensioning, especially with high
hydraulic loading. It should not exceed 2.0 m/h.
4. Organic load <3 kg COD/m^3/day. Higher loading-rates are possible with higher temperature and for easily degradable substrates

Advantages:

1. Resistant to organic and hydraulic shock loads
2. No electrical energy required
3. Greywater can be managed concurrently
4. Can be built and repaired with locally available materials
5. It has long service life
6. No real problems with flies or odours if used correctly
7. High reduction of organics
8. Moderate capital costs, moderate operating costs depending on emptying; can be low cost depending on number of users

Disadvantages:

1. Require constant source of water
2. Effluent requires secondary treatment and/or appropriate discharge
3. Low reduction of pathogens
4. Requires expert design and construction
5. Pre-treatment is required to prevent clogging

Upflow Anaerobic Sludge Blanket Reactor (UASBR)

See Figure 7.

UASB Units

UASB type units are one in which no special media have to be used since the sludge granules themselves act as the 'media' and stay in suspension. A typical arrangement of a UASB type treatment plant for municipal sewage would be as follows:

1. Initial pumping
2. Screening and degritting

Figure 7. Schematic diagram of upflow anaerobic sludge blanket reactor
(Mojiri et al, 2012)

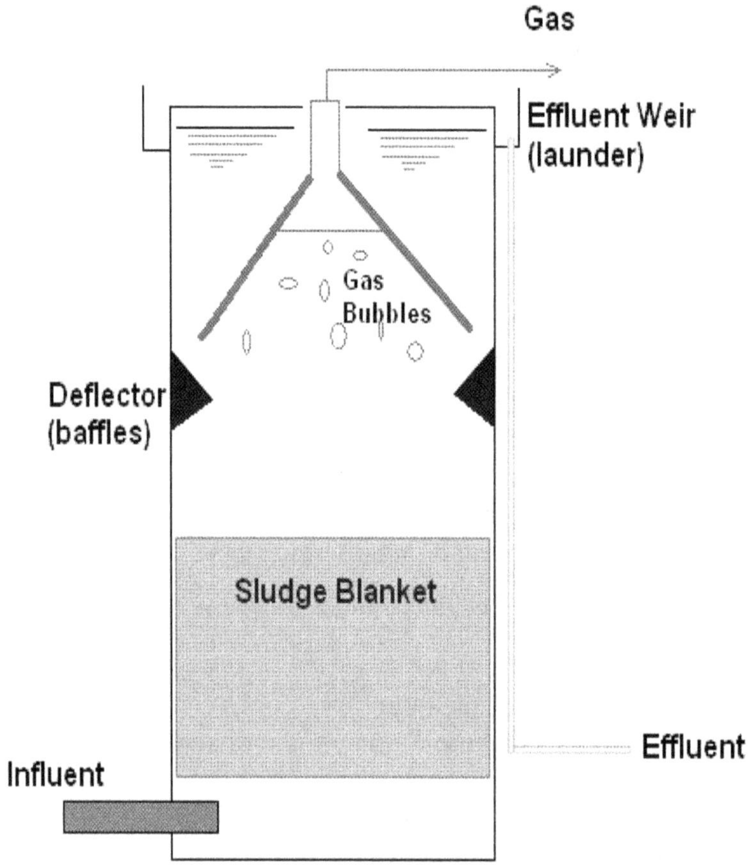

3. Main UASB reactor
4. Gas collection and conversion or conveyance
5. Sludge drying bed
6. Post treatment facility

In the UASBR, the whole waste is passed through the anaerobic reactor in an upflow mode, with a hydraulic retention time (HRT) of only about 8-10 hours at average flow. No prior sedimentation is required. The anaerobic unit does not need to be filled with stones or any other media; the upflowing sewage itself forms millions of small "granules" or particles of sludge which are held in suspension and provide a large surface area on which organic matter can attach and undergo biodegradation. A high solid retention time (SRT) of

Figure 8. General layout of UASB based wastewater treatment plant
(Gupta and Singh, 2015)

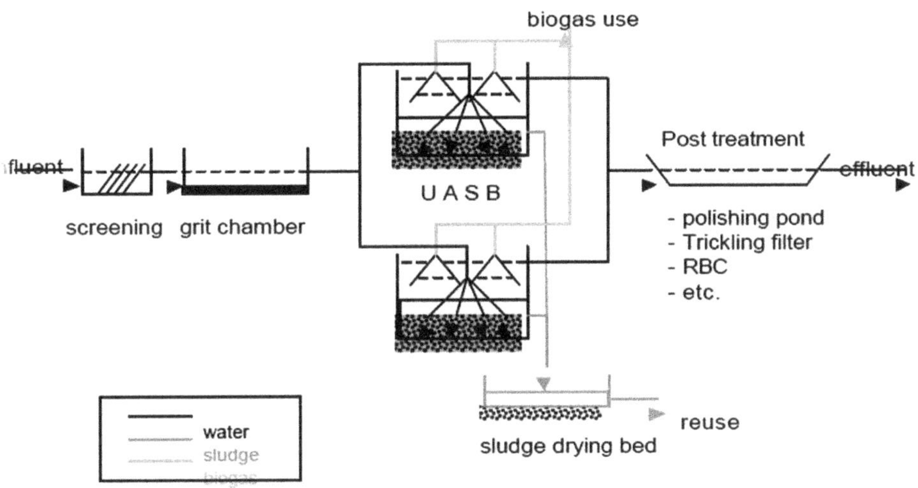

30-50 or more days occurs within the unit. No mixers or aerators are required because the upflow regime and the motion of the gas bubbles allow mixing without mechanical assistance. The gas produced can be collected and used if desired. Anaerobic systems function satisfactorily when temperatures inside the reactor are above 18-20°C. Excess sludge is removed from time to time through a separate pipe and sent to a simple sand bed for drying (Arceivala & Asolekar, 2007).

Anaerobic Processes in the UASB Reactor

There are 4 phases of anaerobic digestion in an UASB reactor

- Hydrolysis, where enzymes excreted by fermentative bacteria convert complex, heavy, un-dissolved materials (proteins, carbohydrates, fats) into less complex, lighter, materials (amino acids, sugars, alcohols).
- Acidogenesis, where dissolved compounds are converted into simple compounds, (volatile fatty-acids, alcohols, lactic acid, CO_2, H_2, NH_3, H_2S) and new cell-matter
- Acetogenesis, where digestion products are converted into acetate, H_2, CO_2, and new cell-matter.
- Methanogenesis, where acetate, hydrogen plus carbonate, formate or methanol are converted into CH_4, CO_2 and new cell-matter.

Specifics of the UASB Reactor

When comparing with other anaerobic reactors, we conclude that the differences, as well as the specifics of an UASB, are existence of granules sludge and internal three-phase GSL device (gas/sludge/liquid separator system).

Granules Sludge

In an UASB reactor, anaerobic sludge has or acquires good sedimentation properties, and is mechanically mixed by the up-flow forces of the incoming wastewater and the gas bubbles being generated in the reactor. For that reason, mechanical mixing can be omitted from an UASB reactor thus reducing capital and maintenance costs. This mixing process also encourages the formation of sludge granules.

The sludge granules have many advantages over conventional sludge flocs:

- Dense compact bio-film
- High settleability (30-80 m/h)
- High mechanical strength
- Balanced microbial community
- Syntrophic partners closely associated

Figure 9. Shape and size of granules sludge

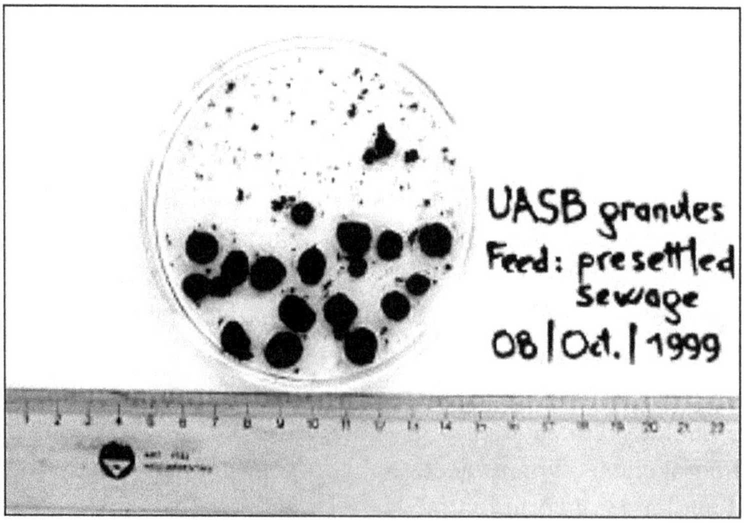

- High methanogenic activity (0.5 to 2.0 g COD/g VSS.d)
- Resistance to toxic shock
- Internal three-phase GSL device: Installed at the top of the tank, the GSL device constitutes an essential part of an UASB reactor with following functions:
- To collect, separate and discharge the biogas formed.
- To reduce liquid turbulences, resulting from the gas production, in the settling compartment.
- To allow sludge particles to separate by sedimentation, flocculation or entrapment in the sludge blanket.
- To limit expansion of the sludge bed in the digester compartment.
- To reduce or prevent the carry-over of sludge particles from the system.

Design Approach

Size of Reactor

Generally, UASBs are considered where temperature in the reactors will be above 20°C. At equilibrium condition, sludge withdrawn has to be equal to sludge produced daily. The sludge produced daily depends on the characteristics of the raw wastewater since it is the sum total of (i) the new VSS produced as a result of BOD removal, the yield coefficient being assumed as 0.1 g VSS/ g BOD removed, (ii) the non-degradable residue of the VSS coming in the inflow assuming 40% of the VSS are degraded and residue is 60%, and (iii) Ash received in the inflow, namely TSS-VSS mg/l. Thus, at steady state conditions,

$$SRT = \frac{Total\ sludge\ present\ in\ reactor, kg}{Sludge\ withdrawn\ per\ day, kg\ /\ d} \tag{10}$$

= 30 to 50 days.

Another parameter is HRT which is given by:

$$HRT = \frac{Reactor\ volume, m^3}{Flow\ rate, m^3\ /\ h} \tag{11}$$

= 8 to 10 h or more at average flow.

The reactor volume has to be so chosen that the desired SRT value is achieved. This is done by solving for HRT from SRT equation assuming (i) depth of reactor (ii) the effective depth of the sludge blanket, and (iii) the average concentration of sludge in the blanket (70 kg/m^3). The full depth of the reactor for treating low BOD municipal sewage is often 4.5 to 5.0 m of which the sludge blanket itself may be 2.0 to 2.5 m depth. For high BOD wastes, the depth of both the sludge blanket and the reactor may have to be increased so that the organic loading on solids may be kept within the prescribed range.

Once the size of the reactor is fixed, the upflow velocity can be determined from

$$Upflow\ velocity\ m\ /\ h = \frac{Reactor\ height}{HRT, h} \tag{12}$$

Using average flow rate one gets the average HRT while the peak flow rate gives the minimum HRT at which minimum exposure to treatment occurs. In order to retain any flocculent sludge in reactor at all times, experience has shown that the upflow velocity should not be more than 0.5 m/h at average flow and not more than 1.2 m/h at peak flow. At higher velocities, carry over of solids might occur and effluent quality may be deteriorated. The feed inlet system is next designed so that the required length and width of the UASB reactor are determined. The settling compartment is formed by the sloping hoods for gas collection. The depth of the compartment is 2.0 to 2.5 m and the surface overflow rate kept at 20 to 28 m^3/m^2-day (1 to 1.2 m/h) at peak flow. The flow velocity through the aperture connecting the reaction zone with the settling compartment is limited to not more than 5 m/h at peak flow. Due attention has to be paid to the geometry of the unit and to its hydraulics to ensure proper working of the "Gas-Liquid-Solid-Separator (GLSS)" the gas collection hood, the incoming flow distribution to get spatial uniformity and the outflowing effluent.

Calculation UASB Tank Base on OLR

If input COD: 5,000 - 15,000 mg/l with Organic loading rate OLR: 4 - 12 kg COD/m^3.d and Hydraulic retention time HRT: 4 - 12 h.

COD treatment efficiency: $E = \frac{(\text{CODinput} - \text{CODoutput})}{\text{CODinput}}$ (13)

In Calculation, Percent of COD removal is 75 - 85%

Organic loading rate $OLR = Q\,(COD_{input} - COD_{output}) \times 10^3$ (14)

Volume of tank, $W = \frac{C \times Q}{OLR} = \frac{(\text{kg COD} / m^3 \times m^3 / h)}{(\text{kg COD} / m^3.h)}$ (15)

C: concentration of COD in wastewater (kg/m^3)
Q: flow rate of wastewater (m^3/h)

H(m) the height of tank can be calculated by:

$$\mathbf{H = H_S + H_{Se}} \quad (16)$$

The height of sludge layer Hs is:

$$\mathbf{Hs = V \times HRT} \quad (17)$$

where Hs: the height of sludge layer area (main reactor) and

H_{se}: the height of sedimentation area
Where V = Velocity of flow 0.6 to 0.9 m/h
HRT = Hydraulic retention time (h)
The height of setting area $H_{Se} \geq 1.2$ m and

The area surface of an UASB tank (m^2):

$$\mathbf{A = \frac{HRT \times Q}{H}} \quad (18)$$

Physical Parameters

A single module can handle 10 to 15 MLD of sewage. For large flows, a number of modules could be provided. Some physical details of a typical UASB reactor module are given in Table 1.

Process Design Parameters

A few process design parameters for UASBs are listed below for municipal sewages with BOD about 200-300 mg/l and temperatures above 20°C.

APPLICATION OF UASB REACTOR FOR WASTEWATER TREATMENT

Suitable Wastewater Characteristics

Granulation of biomass is indicative of successful operation of UASB reactor. Although, acceptable efficiency from the reactor can be obtained when sludge is in flocculent form, existence of granulation sludge configuration offers distinct advantages. The composition of wastewater plays an important role

Table 1. Physical parameters of upflow anaerobic sludge blanket reactor

Reactor Configuration	Rectangular or circular. Rectangular shape is preferred
Depth	It should be 4.5 to 5.0 m for sewage.
Width or Diameter	To limit lengths of inlet laterals to around 10-12 m for facilitating uniform flow distribution and sludge withdrawal.
Length	As necessary.
Inlet Feed	Gravity feed from top (preferred for municipal sewage) or pumped feed from bottom through manifold and laterals (preferred in case of soluble industrial wastewaters).
Sludge Blanket Depth	2 to 2.5 m for sewage. More depth is needed for stronger wastes.
Deflector/GLSS	This is a deflector beam which together with the gas hood (slope 60) forms a "gas-liquid-solid-separator" (GLSS) letting the gas go to the gas collection channel at top, while the liquid rises into the settler compartment and the sludge solids fall back into the sludge compartment. The flow velocity through the aperture connecting the reaction zone with the settling compartment is generally limited to about 5m/h at peak flow.
Settler Compartment	2.0-2.5 m in depth. Surface overflow rate equals 20-28 $m^3/m^2/d$ at peak flow.

Table 2. Process design parameters of upflow anaerobic sludge blanket reactor

HRT	8-10 hours at average flow (minimum 4 hours at peak flow)
SRT	30-50 days or more
Sludge Blanket Concentration (Average)	15-30 kg VSS per m^3. About 70 kg TSS per m^3.
Organic Loading on Sludge Blanket	0.3-1.0 kg COD/kg VSS day (even upto 10 kg COD/ kg VSS day for agro-industrial wastes).
Volumetric Organic Loading	1-3 kg COD/m^3 day for domestic sewage (10-15 kg COD/m^3 day for agro-industrial wastes)
BOD/COD Removal Efficiency	For sewage, it should be 75-85% for BOD. 74-78% for COD.
Inlet Points	Minimum 1 point per 3.7-4.0 m^2 floor area.
Flow Regime	Either constant rate for pumped inflows or typically fluctuating flows for gravity systems
Upflow Velocity	About 0.5 m/h at average flow, or 1.2 m/h at peak flow, whichever is low
Sludge Production	0.15-0.25 kg TS per m^3 sewage treated
Sludge Drying Time	Sludge drying time is seven days (in India)
Gas Production	Theoretical 0.38 m^3/kg COD removed. Actual 0.1-0.3 m^3 per kg COD removed
Gas Utilization	Method of use is optional. 1 m^3 biogas with 75% methane content is equivalent to 1.4 kWh electricity
Nutrients Nitrogen and Phosphorus Removal	5 to 10% only.

in granulation process. Substrates that support granulation are carbohydrates or proteins mainly in soluble, and possibly in colloidal form. Industrial wastewaters from sugar industry, breweries, apple juice, yeast factory, and grape wine satisfy this criterion and give granulation in UASB reactor.

For wastewater containing mainly proteins, granulation proceeds satisfactory; but problems may arise from foaming and protein precipitation under conditions of overloading or low pH less than 6.0. More importantly, they release ammonia upon degradation, which may exert an inhibition effect on microbial activities. High SS concentration in influent can adversely affect granulation and performance of the reactor. The influent SS concentration shall be less than 1 g/L and SS to COD ratio shall be less than 0.5 for successful operation of the reactor.

For wastewater that contains substrates, which do not yield hydrogen in the fermentation process (short chain fatty acids), granulation will not takes place. No granulation in the UASB reactor was reported for acetate only as substrate. For the waste where H_2 generated is preferentially utilized by other organisms such as sulphate reducers, granulation is limited because of reduced amount of H_2 available to the hydrogen trophic methanogens e.g., paper pulping waste. However, the granular yield does not reduce to zero even when SO_4 supplementation is in excess.

Typical industries where UASB reactors are reported to be most successful for wastewater treatment are beet sugar, cane sugar, starch, breweries, dairy, tannery, food processing industries, and paper and pulp. This process is also proved to be feasible for the treatment of domestic wastewater. The feasibility of this process has already been proved for this wastewater in wide COD range, from 500 to above 10,000 mg/L. The treatment is feasible under both mesophilic and thermophilic conditions but temperature above 15°C is essential for proper treatment. When wastewater is mostly in biodegradable form and COD is in the range of 1000 to 5000 mg/L, efficiency of COD removal of 85 to 90% can easily be achieved, with short HRT of 6 to 12 h. When the wastewater is complex, or COD is lower or higher than the above mentioned range, COD removal efficiency of 60 to 80% can still be achieved. Once, the proper start-up of the reactor is achieved with generation of good quality of granular sludge, having good settling characteristics and activity, very high Organic Loading Rates (OLR) greater than 20 kg COD/m^3.d can be applied.

Modes of Operation

UASB reactor is successful for industries listed earlier, where the wastewater coming out of the industry is being continuously treated. It is also successful for wastewater treatment when mode of operation is intermittent. For example, in the case of dairy wastewater treatment, the wastewater is generated only for few hours a day, and not continuously. The process is reported to perform well even under such intermittent mode of operation. Also, this has been experienced that the intermittent operation is useful during initial days of operation to overcome problem of sludge buoying due to poor quality of inoculum used. In case where excessive volatile acids production occurs in UASB reactor, reducing pH lower than 6.5, intermittent mode of operation could be resorted to reduce volatile acids concentration and increase pH in the reactor. UASB reactor is also applicable for the treatment of wastewater from the industries, which are seasonal in origin, like food processing industries. Once, the primary start-up of the reactor is over, with development of good quality of granular sludge, the shutdown of the reactor is possible when the season is over. The reactor put into operation in new season takes very less time (1 to 2 weeks) for this secondary start-up, to restore its COD removal efficiency. For short duration of shut down less than a month reactor can capture its original COD removal efficiency within a week.

Treatment Flow Sheet

The typical flow chart required for UASB type wastewater treatment plant is as follows:

- Screening
- Grit removal (Optional)
- Skimming Tank
- Pumping
- UASB reactor
- Gas collection system
- Post-treatment such as aerobic processes or settling tank, depending on the disposal mode of effluent, and
- Sludge drying beds.

The provision of screens and grit chamber is necessary for the treatment of municipal wastewater as required in conventional wastewater treatment plants. For certain industrial wastewaters provision of screens and grit chamber may not be necessary. When the wastewater contains floating matter such as, oil, grease, soap, pieces of cork and wood, vegetable debris and fruit skins, it is advantageous to have a skimming tank to remove these materials. The presence of oil and grease, if gets adsorbed on the sludge surface, can hinder transport of metabolites and mass transfer, ultimately causing reduction in process efficiency. This may be accomplished in a separate tank or can be combined with primary sedimentation when wastewater also has high suspended solids of inorganic origin.

After the primary treatment, it is required to provide pumping unit to pump the wastewater in upward direction in UASB reactor. Location at which topography of the site suits for utilization of gravity head, choosing appropriate site for UASB reactor may not require pumping. The separate gas collection system can be provided if the gas produced is desired to use for combustion or power generation. Generally, the production of gas in UASB reactor is in the range of 0.25 to 0.35 m3 CH4/kg COD removed. The utilization of biogas for power generation is economical for larger treatment plants.

Post Treatment

The UASB reactor is an efficient process for removal of organic material and suspended solids from sewage or industrial effluents. Particularly, this process is more attractive for treatment of sewage in warm climate. However, the UASB reactor can hardly remove macronutrients (nitrogen and phosphorous), and pathogenic microorganisms are only partially removed. Hence, depending on the final disposal of the effluent quality, post-treatment may be required for removal of suspended solids, organic matter, nutrients, and pathogens present in the raw wastewaters.

After UASB reactor, some form of post treatment is generally desirable depending upon source of effluent discharge. UASB reactor can hardly remove any nitrogen from the wastewater. Hence, effluent from UASB reactor is suitable for irrigation purposes. UASB reactor, when followed by post treatment such as aeration and/or sedimentation, could conveniently achieve irrigation standards. The aeration can be obtained to the effluent flowing through a channel to an irrigation area. Where, the treatment efficiency is adequate to meet the discharge standards, further treatment such as aeration is only necessary to destroy anaerobicity. In such cases, simple cascade type aerator can serve the purpose. In some cases where treatment efficiency is meeting the discharge standards for organic matter but the effluent is high in suspended solids, the use of secondary settling tank becomes essential. When stricter effluent standards have to be met (as for river discharge) some better form of post-treatment may become necessary.

The use of aerobic biological treatment is generally preferred for this polishing treatment. The aerobic process such as bio towers, conventional activated sludge process, or extended aeration can be employed as a second stage treatment. Where the effluent from UASB reactor is expected to be high in nutrient such as nitrogen and phosphorous, the post treatment need to be designed for removal of these nutrients to meet discharge standards for surface water. Shallow oxidation ponds can also be used after UASB reactor for complete treatment of wastewater.

Anaerobic Sequencing Batch Reactor (ASBR)

The anaerobic sequencing batch reactors (ASBRs) are high rate anaerobic treatment processes that operate in the following cyclic steps: feed, reaction, settling, and decantation. The first step involves the addition of substrate

Figure 10. Phases of anaerobic sequencing batch reactor cycle

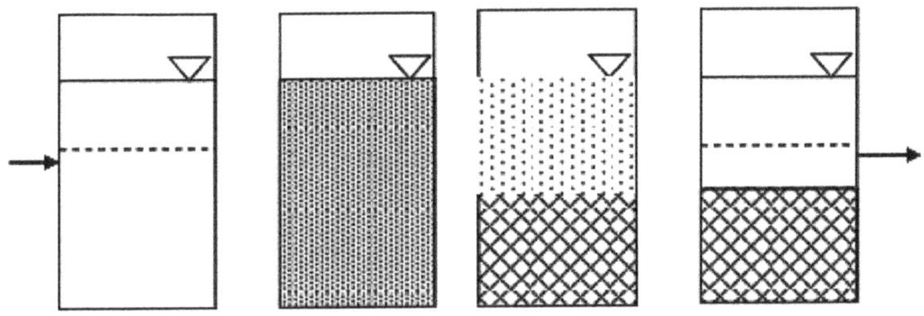

to the reactor where the contents are continuously mixed. The volume of substrate fed depends on a number of factors, including the desired hydraulic retention time (HRT), organic loading, and expected settling characteristics. The conversion of biodegradable organic matter to biogas is achieved during the reaction step. The reactor content is mixed to allow close contact between organics and bacteria. The mixing should, however, be gentle to avoid the disruption of bacterial flocs (a loosely clumped mass of fine particles) formation. The food to microorganism (F/M) ratio is high at the beginning of the reaction phase and the organic conversion into biogas, as predicted by Monod kinetics, is at its maximum. The time required for the reaction step depends on the required effluent quality as well as biomass concentration, substrate characteristics, and type of biomass.

At the end of reaction period, to enable the biomass to floc and settle, mixing of the contents is stopped. The low food to microorganism ratio at the end of reaction phase favours biomass flocculation and settling. During the settling phase, the partial pressure of carbon dioxide above the liquid zone is constant and in equilibrium with dissolved carbon dioxide. As a result, no significant quantity of carbon dioxide is transferred to the headspace, a situation that contributes to the establishment of quiescent settling conditions. The required time for the settling step depends on biomass concentration, temperature, and type of biomass. At the end of the settling step, decantation takes place. During effluent drawdown, microorganisms with poor settling characteristics are also removed from the reactor leaving behind the heavier bacterial flocs. The volume decanted is normally equal to the volume fed during feeding so that the volume is maintained. The discontinuous anaerobic process for wastewater treatment offers several advantages over current anaerobic technologies: there is no short circuit, as in case of fixed bed continuous systems; high biomass concentration; better effluent quality control; absence of primary or secondary settlers; high efficiency for both COD removal and gas production; flexible control, etc., Besides these operating characteristics, some kinetic advantages have also been reported, such as high metabolic activity (Kannan and Singaram, 2012).

Calculations of ASBR

The hydraulic retention time for an ASBR is calculated as follows:

$$\text{HRT} = \frac{V_{ro}}{V_c} \times \text{R} \tag{19}$$

where HRT = Hydraulic retention time, days

V_c = Cycle volume (m^3) R = Cycles per day (day-1)

V_{ro} = Reactor operating volume during react phase (m3)

Solids retention time (SRT) is calculated as follows:

$$\text{SRT} = \frac{V_{ro} \times MLVSS}{CSS \times \frac{V_{sludge}}{t_s} + V_c \times R \times Eff_{VSS}} \tag{20}$$

where SRT = Solids retention time (days)

MLVSS = Mixed liquor volatile suspended solids concentration (kg m-3)
Eff_{VSS} = Effluent volatile suspended solids concentration (kg m-3)
CSS = Sludge volatile suspended solids concentration (kg m-3)
V_{Sludge} = Volume of sludge removed (m3)
t_s = Sludge wasting period (day)

Attached Growth Fluidized Bed Reactor

In this reactor, bacteria are attached to a fine granular material which is fluidized by the upflow of the liquid to be treated, to ensure:

1. A maximum biofilm surface area per unit of volume;
2. An optimum substrate/biofilm contact

This type of reactor allows the most concentrated growth of active bacterial colonies and can therefore accommodate the highest loading. A special biolite medium with a nominal effective size of less than 0.5 mm was selected, based on the strength of the following properties:

- Porous structure (for good biofilm bonding) with high specific surface area;
- Low density (reasonable fluidization velocity - 7 m · h^{-1});
- Resistant to attrition;
- Production conditions, strictly controlled

Design

The reactor can be made of either steel or plastic. Anti-corrosion protection and thermal insulation are usually required. In order to ensure a good fluidization of the medium, a velocity of 5 to 10 m · h^{-1} must be maintained; the process usually requiring effluent recycling. The mixture is evenly distributed

Figure 11. Schematic diagram of fluidized bed reactor

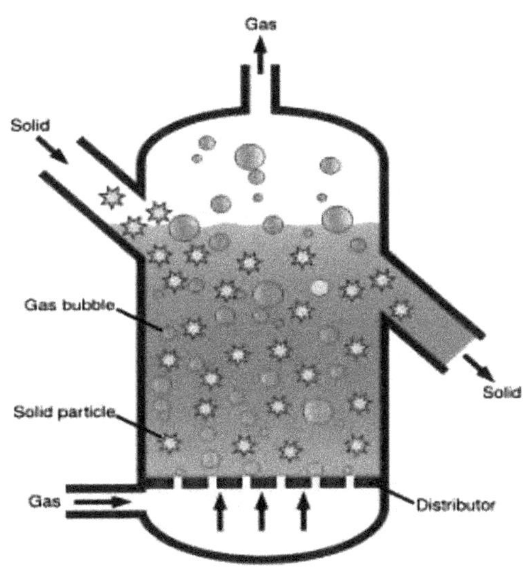

throughout the reactor; a three-phase separator located in the upper section of the reactor is used to release gas and to recover and recycle any entrained biolite (carried by biogas bubbles attached to the film). Gradual colonization of the material can induce excessive segregation, with the consequence of carry-over of light material out of the fluid bed layer. Excess sludge is detached from part of the medium in a high turbulence chamber, enabling the excess biomass to be removed with the treated water.

The main advantages of the process are:

1. No risk of clogging of the support material;
2. Fast start-up;
3. Compact design;
4. Accommodation to high flow rate fluctuations.

Mass Transfer

Principles of Mass Transfer

Mass transfer is a transport of components under a chemical potential gradient. The component moves to the direction of reducing concentration gradient. The transport occurs from a region of higher concentration to lower concentration. Equilibrium is reached when the gradient is zero. The transport or migration of one constituent from a region of higher concentration to that of a lower concentration is known as mass transfer. Mass transfer operations depend on molecules diffusing from one distinct phase to another and are based upon differences in the physico-chemical properties of the molecules, such as vapour pressure or solubility. For interphase mass transfer, there is a concentration gradient between bulk and interface, however under steady state, at interface equilibrium is assumed. Mass transfer operation plays an important role in many industrial processes. A group of operations for separating the components of mixtures is based on the transfer of material from one homogeneous phase to another. These methods is covered by the term mass transfer operations which include techniques like gas absorption and stripping, liquid-liquid extraction, leaching, distillation, humidification, drying, crystallization and number of other separation techniques.

1. **Absorption and Stripping:** Absorption refers to an operation in which the transfer of material is from a gas phase to a liquid phase. A gas is

absorbed by means of liquid in which the solute gas is more or less soluble from its mixture with an inert gas as well as more or less insoluble gas. The liquid is essentially immiscible in the gas phase. The stripping is the separation of gas solute from the liquid phase. The separation of ammonia from an air-ammonia mixture by means of water is a typical example of absorption. The solute is then recovered from the solution by distillation. Depending upon situation, both the fluids (absorbent and absorbing fluid) are reprocessed and/or reused for the operation.

2. **Adsorption and Desorption:** It exploits the ability of transfer of mass from either a gas or a liquid to the surface of a solid. The adsorption is not a true inter-phase mass transfer operation because the fluid adheres to the solid surface instead of dissolving in the solid. A desorption involves the transfer of mass from the solid surface (adsorbents) to the gas or liquid medium (adsorbates). A few operations of practical applications are (a) elimination of toxic gases and deodorization of air, (b) recovery of solvents, (c) removal of ions from solution, as in demineralization of water, (d) fractionation by selective adsorption of gases, vapours from gases, vapors from vapors and liquids from liquids and many other applications.
3. **Extraction:** It refers to a separation of the constituents of a liquid solution by contact with another insoluble liquid. The liquid which is added to the solution to bring about the extraction is known as the solvent. The solution which is to be extracted is called the feed. The solvent-richen product of the operation is called the extract and the residual liquid from which the solute is separated is called the raffinate. The separation of aromatics from kerosene based fuel oils, the production of fuels in the nuclear industry and the separation of penicillin from fermentation mixtures are examples of techniques of extraction. Leaching-is a treatment of a finely divided solid with a liquid. Some examples of leaching operations are oilseed extraction, extraction of sugar beets with hot water and extraction of medicinal compounds from plant roots, leaves and stems.
4. **Distillation:** It is an operation whereby a liquid mixture of miscible and volatile substances is separated into individual components or into group of components by partial vaporization. The separation of a mixture of methanol and water into its components and separation of petroleum crude into gasoline, kerosene, fuel oil and lubricating stock are examples of distillation process. Humidification and dehumidification- In humidification, the enrichment of vapour content in a gas stream takes place by passing the gas over a liquid. The transfer of water from the

liquid phase to the gas phase of a mixture of air and water vapour is a widespread application of humidification. Dehumidification involves the transfer of water vapour from the gas phase to the liquid phase. Removal of water vapour from air by condensation of a cold surface and condensation of carbon tetrachloride out of a stream of nitrogen are examples of dehumidification.

5. **Drying and Evaporation:** Drying refers to removal of relatively small amount of water or other liquid from a solid material whereas evaporation refers to the removal of relatively large amount of water from solutions. In evaporation, the water is removed as vapour at its boiling point. Drying involves the removal of water at temperature below the boiling point by circulating air or some other carrier gas over the material.
6. **Crystallization:** Crystallization is a process of formation of solid from a liquid solution based on difference in solute concentration and its solubility at a certain temperature. In crystallization, transfer of a solute occurs from the liquid solution to a pure solid crystalline phase. When solute concentration becomes higher than its solubility at certain temperature, the solute comes out of the solution and forms a crystal.
7. **Membrane Separation:** This process predominately involves the diffusion of a solute from a liquid or gas through a semi- permeable or microporous membrane to another fluid. Separation of components of the original solution takes place by selectively controlling their passage from one side of membrane to the other. An example of membrane mediated, liquid-liquid separation process is called dialysis. In this process, the microporous membrane allows solutes of low molecular weight in one liquid phase to diffuse readily through it to a second liquid phase by virtue of concentration differences.

Passage of larger colloidal particles through the membrane is difficult. For example, aqueous beet-sugar solutions containing undesired colloidal material are removed by contact with water through a semi-permeable membrane. Sugar and water diffuse through the membrane but not the colloid. If a membrane is placed between a solution and a pure solvent, the solvent diffuses into the solution by osmosis. A reverse pressure, greater than osmotic pressure, is imposed, causing the solvent to flow in the reverse direction, as in the desalination of sea water or brackish water to produce fresh water is called reverse osmosis.

Gas-Liquid Mass Transfer

Gas liquid exchange expressed in two forms:

- Addition of gasses (absorption)
- Removal of gasses (desorption or stripping)

Objectives: Applications

- Addition of O_2 to oxidize Fe & Mn ions & for aerobic biological process such as aeration.
- Addition of Cl_2 for disinfection & removal taste & odour causing compounds such as O_3.
- Removal of H_2S & NH_3 for smell.
- As dissolved CO_2 is added in water, pH decreases thus controlling the scaling.
- Gases in deep well water (CH_4 & CO_2) $\because$ of high pressure & low temperature.
- Removal of volatile organics such as (THM).
- Removal of O_2 for boiler fed water.

Henry's Law

Henry's law is one of the gas laws formulated by William Henry in 1803 and states: "At a constant temperature, the amount of a given gas that dissolves in a given type and volume of liquid is directly proportional to the partial pressure of that gas in equilibrium with that liquid." An equivalent way of stating the law is that the solubility of a gas in a liquid is directly proportional to the partial pressure of the gas above the liquid:

$$C_s \propto p$$

where C_s- Saturation concentration of any gas

p- Partial pressure of any gas in contact with liquid

$$C_s = K_H \cdot p$$

or ml of gas dissolved in 1 L of liquid, K_H = Henry's constant (ml/l.atm STP= standard temperature pressure (T= 0°C, P=1 atm or 760mm Hg ∵ 1 mole of gas occupies 22.4L of volume at STP

Gas–liquid reactions and contacting operations are exceedingly common in the processing of chemicals. Reactions such as oxidations, chlorinations, hydrogenations, and polymerization of alkenes require efficient contacting of gases and liquids. Gas–liquid mass transfer without reaction occurs typically in separation processes such as distillation and absorption. Biotechnology processes for culturing microorganisms often require transfer of oxygen and carbon dioxide. Gas–liquid mass transfer can be especially difficult in viscous fluids such as fermentation broths and polymer solutions.

Generally, gases are sparingly soluble in liquids. Consequently, the supply of a dissolved gas in a reactor is rapidly exhausted by the reaction and the dissolved gas must be replenished continuously by bubbling. The transport route of a soluble gas from the gas to the liquid phase is schematically illustrated in Figure 12. The gas may be imagined as passing through stagnant gas and liquid films on either side of the gas–liquid interface. The resistance to transfer is localized within the stagnant films and the gas–liquid interface itself is assumed to offer no resistance to mass transfer. Mass transfer in the films occurs solely by diffusion; therefore, at steady state, linear concentration profiles exist in the films (Fig. 12). The transport flux (J) of the diffusing

Box 4.

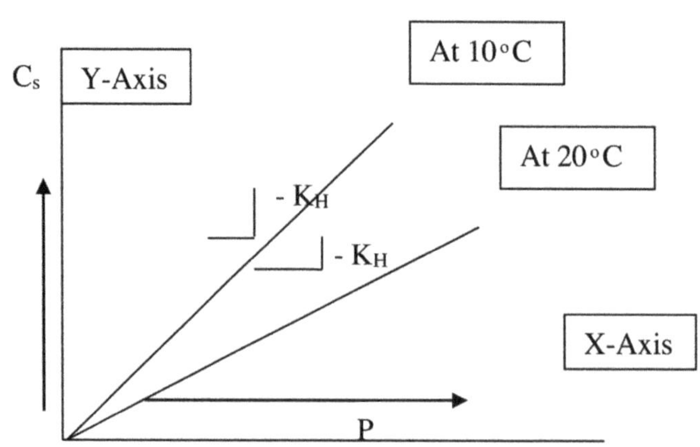

$Y = mx + C$

$C_s = K_H . p + 0$ ∴ intercept =0 & slope = K_H

Table 3. K_H values of gases at different temperature

Gas Type	K_H (ml/L) at 1 Atmosphere and at STP			
	0°	10°	20°	30°
O_2	49.3	38.4	31.4	26.7
N_2	23	-	15.5	-
NH_3	1300	910	711	-
CO_2	1710	-	878	-
O_3	641	-	368	-
H_2S	4690	-	2670	-
CH_4	55.6	41.8	33.1	-

- $NH_{3,}$ CO_2 & H_2S are highly soluble
- Others slightly soluble

Figure 12. Schematic diagram of gas-liquid mass transfer

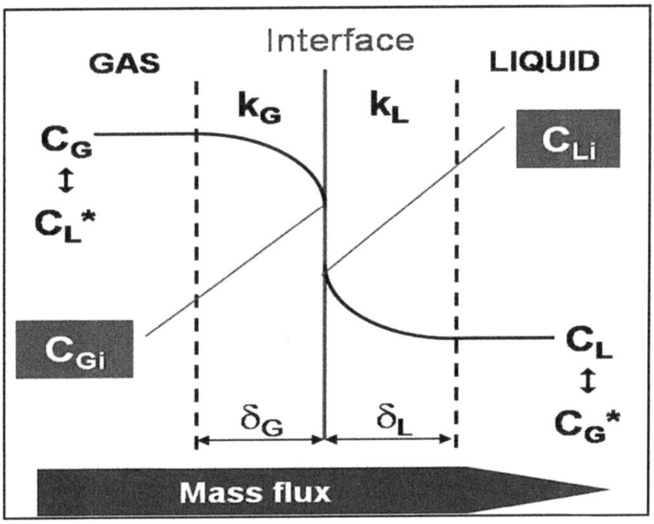

species is related to the concentration gradient (DC) in the film and to the film thickness (d) as follows:

$$J = \frac{D}{\delta} \cdot C \tag{21}$$

In equation 21, D is the diffusivity of the transferring component in the film. The ratio D/d is known as the mass transfer coefficient, k. For an ideal gas, the thickness δ_g of thc laminar gas film at the gas–liquid interface is rclated with diffusivity (D_g) and pressure (P_g), as follows:

$$\delta_g = \frac{P_g D_g}{k_g T R_g} \tag{22}$$

where k_g (kmol/sm^2) is the gas-film mass transfer coefficient. At steady state, the flux of the diffusing component—or the rate of transfer per unit cross-sectional area—is the same through the gas and the liquid films; thus, equation 3 may be written for each of the two films:

$$J = k_G (C_G - C_{Gi}) . \tag{23}$$

$$J = k_L (C_{Li} - C_L)$$

where k_G and k_L are the gas and the liquid film mass transfer coefficients, respectively. The overall mass flux from the gas to the liquid phase may be written as follows:

$$J = K_L (C^* - C_L)$$

where C* is the saturation concentration (ie, the maximum possible value) of the diffusing component in the liquid in equilibrium with the gas phase and K_L is the overall mass transfer coefficient based on the liquid film. The saturation concentration C* in the liquid is related to the gas-phase concentration (C_G) of the diffusing component by Henry's law:

$$C_G = HC^*$$

where H is the dimensionless Henry's constant.

By using above equations and the knowledge that $C_{Gi} = H.\ C_{Li}$ (ie, equilibrium exists at the interface because the interface offers no resistance to mass transfer), the various mass transfer coefficients can be shown to be related as follows

$$\frac{1}{K_L} = \frac{1}{k_L} + \frac{1}{Hk_G} \tag{24}$$

For a sparingly soluble gas such as oxygen, H is much greater than unity (H ~ 30 for oxygen in water); furthermore, the coefficient k_G is typically > k_L because the gas-phase diffusivities are vastly greater than those in liquids (eg, $D_{oxygen/air} = ¼ 10^4 D_{oxygen/water}$ at 208°C).

Liquid Solid Mass Transfer

Mass transfer to or from suspended solids is important in many processing situations. Solid–liquid (or gas) mass transfer may be rate limiting in heterogeneous catalysis, during dissolution of solids, adsorption, as well as in other cases. The rate of mass transfer to or from the suspended particle depends on the solid– liquid mass transfer coefficient (k_L), the total solid–liquid interfacial area (As), and the concentration driving force; thus,

$$V_L \frac{dC_L}{dt} = k_L A_s (C^*-C_L) \tag{25}$$

where C_L is the concentration of the transferring component in the liquid at time t, C* is the saturation concentration (or solubility) of the transferring component, and V_L is the volume of the suspending liquid. Above equation is written for transfer from the solid to the liquid. By analogy with the film model of mass transfer, the coefficient k_L conceptually equals the diffusivity of the solute divided by the thickness of the stagnant liquid film at the solid–liquid interface; k_L is needed for quantifying the rate of mass transfer. Methods for estimating the solid–liquid or solid–gas (also gas–liquid if liquid is adsorbed as a very thin film on the surface of the solid) mass transfer coefficient k_L in various situations are discussed in a later section of this monograph.

Mass Transport Processes

Processes which move pollutants and other compounds through the air, surface water, or subsurface environment or through engineered systems (for example, treatment reactors) are of particular interest to environmental engineers and scientists. Pollutant transport acts to move pollutants from the location at which they are generated, resulting in impacts which can be distant from the pollution source. On the other hand, some pollutants, such as sewage sludge, can be degraded in the environment if they are sufficiently dilute. For these pollutants, slow transport---slow dilution---can result in excessively high pollutant concentrations, with resulting increased adverse impacts.

In this section, we discuss the processes which distribute pollutants in the environment. The goals of this discussion are twofold: to provide an understanding of the processes which cause pollutant transport, and to present and apply the mathematical formulas used to calculate pollutant fluxes.

Advection and Diffusion

Transport processes in the environment may be divided into two categories: advection and diffusion. Advection refers to transport with the mean fluid flow. For example, if the wind is blowing towards the east, advection will carry any pollutants present in the atmosphere toward the east. Similarly, if a bag of dye is emptied into the center of a river, advection will carry the resulting spot of dye downstream. In contrast, diffusion refers to the transport of compounds through the action of random motions. Diffusion works to eliminate sharp discontinuities in concentration and results in smoother, flatter concentration profiles. Advective and diffusive processes can usually be considered independently. In the example of a spot of dye in a river, while advection moves the center of mass of the dye downstream, diffusion spreads out the concentrated spot of dye to a larger, less concentrated region.

Mass Flux Density

We use the term mass flux ($\dot{m}$, with units of mass/time) when we calculate the rates at which mass gets transported into and out of a control volume in a mass balances. Because mass balance calculations are always made with reference to a specific control volume, it was clear that this value referred to the rate at which mass was transported across the boundary of the control volume. In our calculations of advective and diffusive fluxes, however, we will not restrict ourselves to a specific, well-defined control volume. Instead, we will calculate the flux density across an imaginary plane oriented perpendicular to the direction of mass transfer. We will use the symbol 'J' to represent the flux density. **J** represents the mass flux density, expressed as the rate per unit area at which mass is transported across an imaginary plane. J has units of [M]/[LzT].

The total mass flux across a boundary ($\dot{m}$) can be calculated from the flux density simply by multiplying J by the area of the boundary:

$$\dot{m} = J.A \tag{26}$$

In the following sections, we will consider the flux density which results from advection and from diffusion. The symbol J will be used to represent the flux density in each case, whether the flux is a result of advection, diffusion, or a combination of both processes.

Calculation of the Advective Flux

The advective flux refers to the movement of a compound along with flowing air or water. The advective flux density depends simply on concentration and flow velocity.

$$J = C.v \tag{27}$$

The fluid velocity, **v**, is a vector quantity---it has both magnitude and direction, and the flux **J** refers to the movement of pollutant mass in the same direction as the fluid flow. In this course, we will generally define our coordinate system so that the x-axis is oriented in the direction of fluid flow. In this case, the flux **J** will reflect a flux in the x-direction, and we will generally ignore the fact that **J** is really a vector.

Example 1: Calculate the average flux density J of phosphorus downstream of the sewage pipe. The cross-sectional area of the river is $30m^2$ with flow rate of $25m^3/s$?

Solution: In example 1, we found the following conditions downstream of the spot where a sewage pipe added effluent to a river with volumetric flow rate and as phosphorus concentration at downstream as $Q = 2525$ m^3/s, $C_{d/s} = 0.20$ mg/l.

The average river velocity calculated as:

$$v = \frac{Q}{A} = \frac{25m^3 / s}{30m^2} = 0.83m / sec$$

Using the definition of flux density (equation 2), we find:

$$J = C.v \; J = \left(0.20 \frac{mg}{l}\right) \frac{10^3 l}{m^3} X 0.83 m / sec = 166 \text{ mg/m}^2 \bullet \text{s Ans.}$$

Diffusion

Diffusion results from random motions of two types: the random motion of molecules in a fluid, and the random eddies which arise in turbulent flow. Diffusion from the random molecular motion is termed molecular diffusion; diffusion which results from turbulent eddies is called turbulent diffusion or eddy diffusion. We will compare these two types of diffusion below. First, however, we consider why diffusion occurs.

Fick's Law of Diffusion

In this section, we will derive Fick's Law, the equation which is used to calculate the diffusive flux density, by analyzing the results of random motion of a hypothetical box of molecules. The purpose of this derivation is to provide a qualitative and intuitive understanding of the reason that diffusion occurs, and the derivation itself is useful only for that purpose. In problems where it is necessary to actually calculate the diffusive flux, we will normally start at the end of this derivation---that is with the Fick's Law equation. Consider a box which is initially divided into two parts, as shown in Figure 8. Each side of the box has a height and depth of 1 unit and a width of length ΔX. Initially, the left portion of the box is filled with 10 molecules of gas x and the right side is filled with 20 molecules of gas y, as shown in the top half of Figure 13. We now remove the divider and observe what happens.

As molecules are never stationary. Therefore all of the molecules in our box are constantly moving around, and at any moment they have some probability of crossing the imaginary line at the center of the box. We will check the box and count the molecules on each side every Δt seconds; we will call the probability that a molecule crosses the central line during the period between observations k. Let's say for now that k=0.20. So the first time we check the box, after a period Δt, 20% of the molecules that were originally on the left will have moved to the right, and 20% of the molecules that were originally on the right will have moved to the left.

When we count the molecules on each side, then, we will find the situation shown in the bottom of Figure 13, with 8 "x" molecules remaining on the left and 2 on the right, and 16 "y" molecules remaining on the right,

Figure 13. Diffusion

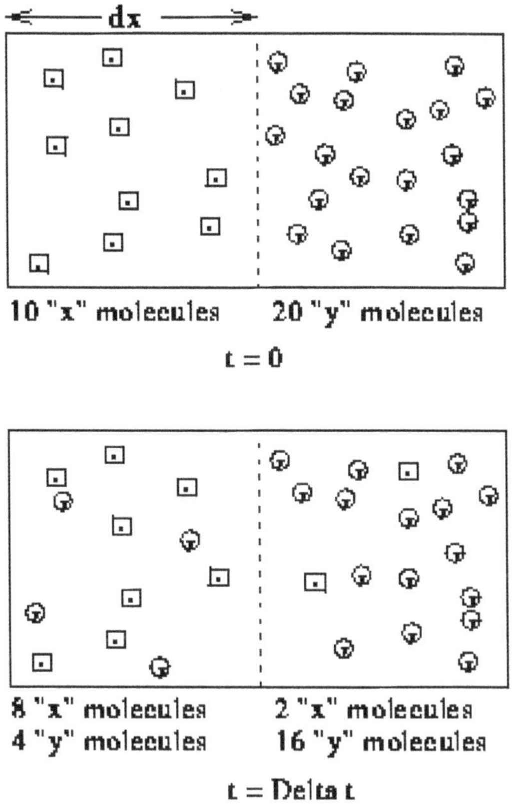

4 having moved to the left. If we assume that the mass of each molecule is 1 unit, we can calculate concentrations in units of mass/volume, and find that the concentration differences between the two boxes have been reduced from 10 to 6 for molecule "x", and from 20 to 16 for molecule "y." This is a fundamental result of diffusion---that concentration differences are reduced. We also observe that the movement of molecule "x" is essentially independent from that of "y", so the diffusion of each molecule can be considered as a separate problem. That is, we don't have to worry about molecule "x" when we are calculating the diffusion of molecule "y", or vice versa.

So diffusion moves mass from regions of higher concentration to regions of lower concentration, and if left to continue indefinitely, it would result in equal concentrations on both sides of the box. We now need to find the flux density J which diffusion causes. For this calculation, we will again use the situation shown in Figure 13, with a probability of any molecule crossing the central boundary during a period Δ t equal to k. Since each molecule can be considered independently, we will analyze the movement of a single molecule type, say molecule "y."

Let M_L be the total mass of molecule "y" in the left half of the box, and M_R equal the mass in the right half. Since our box has unit height and depth, the area perpendicular to the direction of diffusion is one square unit. Thus, the flux density---the flux per unit area---is just equal to the rate of mass transfer across the boundary. The amount of mass transferred from the left to right in a single time step is equal to kM_L, while the amount transferred from right to left during the same period is kM_R. Thus, the *rate* of mass flux from left to right across the boundary is equal to (kM_L-kM_R) divided by Δt, or

$$J = \frac{k}{\Delta t}\left(M_L - M_R\right) \tag{28}$$

Since it is more convenient to work with concentrations than with total mass values, we need to convert this equation to concentration units. The concentration in each half of the box is given by

$$C_L = \frac{M_L}{\Delta x.\left(height\right).\left(depth\right)} = \frac{M_L}{\Delta x} \text{(height and depth both are unity),} \tag{29}$$

and

$$C_R = \frac{M_R}{\Delta x} \tag{30}$$

Thus, substituting CΔxfor the mass in each half of the box, the flux density is equal to

$$J = \frac{k}{\Delta t}\left(C_L\Delta x - C_R\Delta x\right) = \left(\frac{k}{\Delta t}\right)\Delta x\left(C_L - C_R\right) \tag{31}$$

Finally, we note that as

$$\Delta x \to 0,\ \left(C_L - C_R\right) / \Delta x \to dC / dx$$

and therefore

$$J = -\frac{k}{\Delta t}\left(\Delta x\right)^2 \frac{dC}{dx} \qquad (32)$$

(Note that the negative sign in this equation is simply a result of the convention that flux is positive when it flows from left to right, but the derivative is positive when concentration *increases* toward the right.) This equation states that the flux of mass across an imaginary boundary is proportional to the concentration gradient at the boundary. Since the resulting flux cannot depend on arbitrary values of Δt or Δx, the product $\frac{k}{\Delta t}\left(\Delta x\right)^2$ must be constant. This product is the value we call the *Diffusion Coefficient*, D. Thus, we obtain Fick's Law:

$$J = -D\frac{dC}{dx} \qquad (33)$$

The diffusion coefficient has the same units as $\frac{k}{\Delta t}\left(\Delta x\right)^2$. Since k is a probability, and thus has no units, the units of D must be (length)z /(time). Diffusion coefficients are commonly reported in cm^2/s.

This form of equation will also appear later when we discuss Darcy's Law, which governs the rate at which groundwater flows through soil pores. The same equation also governs heat transfer, if we replace the concentration gradient with a temperature gradient.

Molecular Diffusion

The molecules-in-a-box analysis used above is essentially an analysis of molecular diffusion. Purely molecular diffusion is relatively slow. Typical values of the diffusion coefficient D are approximately 10^{-2}-10^{-1} cm^2/s for gases, and much lower, around 10^{-5} cm^2/s for liquids. The difference in diffusion coefficient between gases and liquids is understandable if we consider that gas molecules are free to move much greater distances before being stopped by "bumping" into another molecule. The diffusion coefficient also varies with temperature and the molecular weight of the diffusing molecule. This is because the average speed of the random molecular motions is dependent

Table 4. Selected molecular diffusion coefficients in water and air

Compound	Temp. (°C)	Diffusion Coefficient (cm^2/s)
Methanol in H_2O	15	1.26×10^{-5}
Ethanol in H_2O	15	1.00×10^{-5}
Acetic Acid in H_2O	20	1.19×10^{-5}
Ethyl benzene in H_2O	20	8.1×10^{-5}
CO_2 in air	20	0.151

on the kinetic energy of the molecules. As heat is added to a material and temperature increases, the thermal energy is converted to random kinetic energy of the molecules, and the molecules move faster. This results in an increase in the diffusion coefficient with increasing temperature. However, if we compare molecules of differing molecular weight, we find that at a given temperature a heavier molecule moves more slowly, and thus the diffusion coefficient decreases with increasing molecular weight.

Q-1 Gasoline-contaminated groundwater has been transported under a house from a nearby gas station. Two meters below the dirt floor of the house's basement, the concentration of hydrocarbon vapors in the airspace within the soil is 3 x 10^{-8}g/cm^3. Estimate the flux density of gasoline vapor transported into the basement by molecular diffusion. The diffusion coefficient for gasoline vapor in the air space within the soil is equal to $10^{-2}cm^2/s$. Assume that the basement is well-ventilated so that the concentration of gasoline in the basement is very small in comparison to the concentration in the soil.

Sol. To calculate the flux density, we use Fick's Law, assuming that the gradient of concentration with height is linear over the 2 m depth. We will not be careful about the sign in the equation, however, since we know that the diffusive flux will be from the ground, where concentration is higher, into the basement, where the concentration is lower.

$$J = -D\frac{dC}{dz} = D\frac{\Delta C}{\Delta z}$$

$$J = \left(10^{-2}cm^2 / s\right).\frac{(3\times10^{-8})g / cm^3}{200\,cm}$$

$$J = \left(10^{-2}cm^2 / s\right)\cdot 1.5\times10^{-10}g / cm^4$$

$$J = \left(1.5 \times 10^{-12} g\ /cm^2 s\right).$$

The calculated flux density of 1.5×10^{-12} g/cm^{2}s can be used to calculate the total mass flux of gasoline into the basement using equation. If the basement floor has an area of 100 m^2, then the total flux into the basement is

$$\dot{m} = J \times A$$

$$= 1.5 \times 10^{-12} g\ /\ cm^2 s \times 100 m^2 10^4 \frac{cm^2}{m^2} \frac{3600s}{hr} \frac{24hr}{day}$$

$$= 0.13 g\ /\ day$$

Turbulent Diffusion

In turbulent diffusion, mass is transferred through the mixing of turbulent eddies within the fluid. This is fundamentally different from the processes which determine molecular diffusion---in turbulent diffusion, it is the random motion of the fluid that does the mixing, while in molecular diffusion it is the random motion of the pollutant molecules that is important. Earlier in this section, we used an example in which a spot of dye was dropped into the center of a river. If we follow the center of the dye spot down the river, we would see that spreading out of the spot by molecular diffusion would occur (slowly) as a result of the random motion of dye molecules across the edges of the spot. Turbulent diffusion would occur (much more quickly), as a result of eddies in the river mixing clean water from outside the spot with dye-colored water within the spot.

To indicate this difference in causation, the diffusion coefficient for turbulent diffusion is often referred to as the eddy diffusion coefficient. The value of the eddy diffusion coefficient depends on the properties of the fluid flow, rather than on the properties of the pollutant molecule we are interested in. Most important is the flow velocity---turbulence is only present at flow velocities above a critical level, and the degree of turbulence is correlated with velocity. (More precisely, the presence or absence of turbulence depends on the Reynolds Number, a non-dimensional number which depends on velocity, width of the river or pipe, and the viscosity of the fluid.) In addition, the degree of turbulence depends on the material over which the flow is occurring, so that flow over bumpy surfaces will be more turbulent than flow over a smooth surface, and the increased turbulence will cause more rapid mixing.

Finally, the value of the eddy diffusion coefficient depends to some extent on the size scale of the problem we are considering. This is best illustrated by an example. Figure 14 shows three examples of the mixing of an isolated puff of pollutant (or spot of dye) in a turbulent flow. In Fig. 14 the size of the pollutant puff is large compared to the size of the turbulent eddies. The result is that turbulent diffusion is slow. The opposite extreme is shown in Figure 14 b, where the size of the puff is very small compared to the turbulent eddies. In this case, the entire puff is moved along with the fluid eddies. The result is not diffusion at all---we would normally call this advection since the puff is being "blown" along intact. The third example (Figure 14 c) shows the intermediate situation. Here, the size of the puff is comparable to the size of the turbulent eddies, and the puff is rapidly stretched out and mixed with the surrounding fluid. In this case, the eddy diffusion coefficient would be rather large. In the real world, of course, turbulent eddies of all sizes are present simultaneously. Therefore, any given case of turbulent diffusion will be a

Figure 14. Scales of turbulence

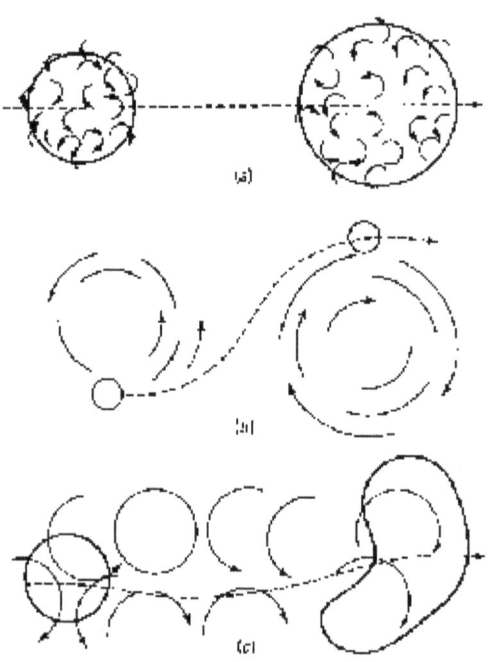

mixture of the three situations shown in Figure 14, and only a single eddy diffusion coefficient would be used.

Fick's Law applies to turbulent diffusion just as it does to molecular diffusion. Thus, flux density calculations are the same for both processes; only the magnitude of the diffusion coefficient is different.

Mechanical Dispersion

The final diffusion-like process that we will consider is similar to turbulence, in that it is a result of variations in the movement of the water (or air) which carries our pollutant. In *mechanical dispersion*, these variations are the result of (a) variations in the flow pathways taken by different fluid parcels that originate in the nearby locations near one another, or (b) variations in the speed at which fluid travels in different regions.

Two fluid parcels starting near each other at locations B and C are dispersed to locations farther (B' and C') during transport through the soil pores, while parcels from A and B are brought closer together, resulting in mixing of water from the two regions

Dispersion in groundwater flow provides a good example of the first process. Figure 15 shows a magnified depiction of the pores within a soil sample, through which groundwater flows. (Note that, as shown in Figure 15, groundwater movement is not a result of underground rivers or creeks, but rather is caused by the flow of water through the pores of the soil, sand, or other material underground.) Because transport through the soil is limited to the pores between soil particles, each fluid particle takes a convoluted path through the soil and, as it is transported horizontally with the mean flow, it is displaced vertically a distance that depends on the exact flow path it took. The great variety of possible flow paths results in a random displacement

Figure 15. The process of mechanical dispersion in groundwater flow

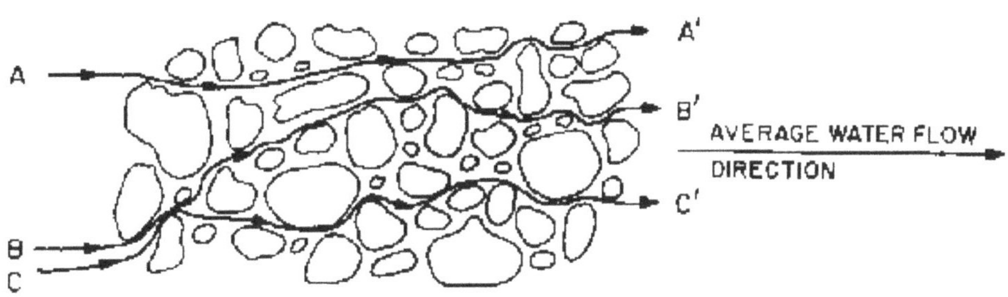

in the directions perpendicular to the mean flow path. Thus, a spot of dye introduced into the groundwater flow between points B and C in Figure 15 would be spread out or dispersed into the region between points B_f and C_f as it flows through the soil.

The second type of mechanical dispersion results from differences in flow speed and is only important in non-steady state problems, such as the accidental, sudden release of a pollutant into a flowing stream. Anywhere that a flowing fluid contacts a stationary object, the speed at which the fluid moves will be slower near the object. For example, the speed of water flowing down a river is fastest in the center of a river and can be very slow near the edges. Thus, if a line of dye were somehow laid across the river at one point, it would be stretched out as it flowed down the river, with the center part of the line moving faster than the edges. This type of dispersion spreads things out in the longitudinal direction---in the direction of flow---in contrast to diffusion and dispersion in groundwater, which spread things out in the direction perpendicular to the direction of mean flow.

NUMERICAL

- Q-1 Design a UASBR System with (i) the size and dimension of system (ii) HRT (iii) SRT(θ_C) (iv)Average VSS concentration of biomass zone of reactor (v) Gas production (vi) Energy available from methane (vii) Alkalinity required for wastewater with the characteristics given above to achieve greater than 90% soluble COD removal.
- Assume 50% of the influent pCOD and VSS is degraded
- Assume 90% of the influent sulfate is reduced biologically and the effluent VSS concentration is 150g/m^3
- Assume reactor effectiveness factor = 85%
- Height of gas chamber = 2.5m
- Organic Loading Rate = 15kg/COD/m^{3}d

Sol. Nominal Liquid Volume $V_N = \dfrac{QS_0}{OLR}$ (34)

Q = 1000m^3/d

S_o= 2000gm/m^3 = 2kg/m^3

OLR= 12-20kgCOD/m^3/d

Assume OLR= 15kgCOD/m^3/d

$$V_N = \frac{1000 \times 2}{15} \frac{m^3}{d} \times \frac{kg}{m^3} \times \frac{m^3 d}{kg}$$

V_N = 135m^3 (approx)

Gross Liquid Volume, $V_L = \frac{V_N}{E} = \frac{135}{O.85}$ m^3= 160 m^3 (approx) (35)

$$\text{Reactor Area} = \frac{Q}{V} = \frac{1000\text{m}^3 / \text{d}}{1.5 m / h} = 666.67 \frac{m^3 h}{d} \quad (36)$$

$$= \frac{666.67}{24} m^2 = 27.8\text{m}^2$$

$$d = \sqrt{\frac{A \times 4}{\pi}} = 6 \text{ m (approx)} \quad (37)$$

$$H_T = H_G + H_L \quad (39)$$

$$H_L = \frac{V_L}{A} = \frac{160}{27.8} = 5.75\text{m}$$

H_T = 5.75 + 2.5 =8.25m

$$\text{HRT} = \frac{V_L}{Q} = \frac{160}{1000} \frac{m^3 d}{m^3} = 3.84\text{hr}$$

Range of HRT should be 4 to 8hrs

Therefore we have to calculate with different OLR values

Assume OLR = 10kgCOD/m^3/d

Again Calculate V_N,

$$V_N = \frac{QS_0}{OLR} = \frac{1000 \times 2}{10} = 200\ \text{m}^3$$

$$V_L = \frac{V_N}{E} = \frac{200}{0.85} = 235.29\ \text{m}^3$$

From eq.27 Reactor Area= 27.8m^2

From eq.28 d = 6m

$$H_T = H_G + H_L \qquad (40)$$

$$H_L = \frac{V_L}{A} = \frac{235.29}{27.8} = 8.5\text{m}$$

$$H_T = 8.5 + 2.5 = 11\text{m}$$

$$\text{HRT} = \frac{V_L}{Q} = \frac{235.29}{1000}\frac{m^3 d}{m^3} = 5.64\ \text{hr}$$

(Now the value is in range 4 to 8hrs, therefore it is good for design)

Now sludge retention time should be kept between 50-100 days,

QX_c = solids wasted per day

Data needed:-

1. The effluent soluble COD concentration at 90% COD removal is

$S = 0.1 \times 2000 g/m^3 = 200 g/m^3$

2. The effluent nbVSS concentration given that 50% of the influent VSS is degraded is:

$nbVSS = 0.5 \times 150\ g/m^3 = 75\ g/m^3$

3. The pCOD degraded is $0.5(2300\text{-}2000) g/m^3$
4. Total degradable influent COD is, S_o

$S_o = (2000+150) g/m^3 = 2150\ g/m^3$

$Y = 0.08 gVSS/gCOD$

$k_d = 0.03\ gVSS/gVSS\text{-}d$

$\mu_m = 0.25\ gVSS/gVSS\text{-}d$

$f_d = 0.15g$ VSS cell debris/g VSS biomass decay

Methane production at $35°C = 0.40 LCH_4/gCOD$

$$QX_C = \frac{Q(Y)(S_0 - s)}{1+(K_d)SRT} + \frac{f_d(K_d)Q(Y)(S_0 - s)SRT}{1+(K_d)SRT} + Q_{(nbvss)} \qquad (41)$$

$Q_{(nbvss)} = Q_{\text{non biodegradable volatile solids substances which can be inorganic in nature}}$

$$QX_C = 1000\frac{m^3}{d} \times 150\frac{g}{m^3} = 1{,}50{,}000 g/d$$

$$\frac{Q(Y)(S_0 - s)}{1+(K_d)SRT} = \frac{\left(1000\frac{m^3}{d}\right)\left(0.08\frac{\text{gVSS}}{\text{gCOD}}\right)[(2150-200)\text{g}/\text{m}^3}{1+\left(0.03\frac{gVSS}{gVSS-d}\right)(SRT)}$$

$$\frac{f_d\left(K_d\right)Q\left(Y\right)\left(S_0 - s\right)SRT}{1+\left(K_d\right)SRT} = \frac{0.15 \times 0.03 \times 100000 \times 0.08 \times 1950 \times SRT}{1+\left(0.03\right)SRT}$$

$$= \frac{702SRT}{1+\left(0.03\right)SRT}$$

$$Q_{(nbvss)} = 1000\frac{m^3}{d} \times 75\frac{g}{m^3} = 75000\frac{g}{d}$$

$$150000 = \frac{156000}{1+0.03SRT} + \frac{702SRT}{1+0.03SRT} + 75000$$

Hence SRT = 54d

Now estimate the effluent soluble COD at an SRT of 54d at 30°C

$$S = \frac{K_s(1+\left(K_d\right)\theta_c}{\theta_c\left(YK - K_d\right) - 1},\ K = \frac{\mu_m}{Y} = \frac{0.25}{0.08} = 3.125\frac{gCOD}{gVSS - d}$$

$$S = \frac{\left(360\frac{mg}{L}\right)\left[1+0.03\times 54\right]}{54\left(0.08\times 3.125 - 0.03\right) - 1} = 86.69\text{mg/L}$$

Hence computed SRT value is adequate

$$\frac{Influent\,COD}{Effluent\,COD} = \frac{86.69}{2000} = 0.043 = 4.3\%\text{S}$$

Because 4.3% is less than 10%, the process SRT is adequate

#Determine average X_{TSS} concentration in biomass zone of reactor

$$(\text{SRT})\,\theta_c = \frac{V\left(X_{TSS}\right)}{\left(Q - Q_w\right)X_e + Q_w X_R} = \frac{\text{MLSS in the system}}{\text{MLSS leaving the system}}$$

Assume $Q_W = 0$, $\theta_c = \dfrac{V\left(X_{TSS}\right)}{QX_e}$ and $X_{TSS} = \dfrac{\theta_c QX_e}{V}$

$$X_{TSS} = (54\text{d})(1000\frac{m^3}{d})(150\frac{g}{m^3})\frac{1}{200m^3} = 40.5\text{kg}/m^3$$

Determining Methane gas production and energy produced

COD degraded = $(2150\text{-}200)\dfrac{g}{m^3} = 1950\dfrac{g}{m^3}$

COD removed with sulfate(0.67g COD removed/gSO_4)

$$COD_{SR} = 0.90(200\frac{\text{gSO4}}{m^3})(0.67\frac{gCOD}{gS0_4}) = 120.6\frac{g}{m^3}$$

COD used by Methanogenic bacteria

$$COD_{MB} = (1950\text{-}120.6)1000\frac{m^3}{d}\times\frac{g}{m^3} = 1{,}829{,}400 \text{ g/d}$$

Methane production rate at 30^{0}C = $(0.4\dfrac{L}{g})(\dfrac{273.15+30}{273.15+35}) = 0.3935\dfrac{L}{g}$

Amount of methane produced/day = $0.3935(\dfrac{L}{g})(1829400\dfrac{g}{d})$

$$= 719868.9\frac{L}{d} = 719.8\frac{m^3}{d}$$

Total gas volume produced (use 65% methane) $= 719.8\dfrac{m^3}{d}\times 0.65 = 1107.38$

$$\frac{m^3}{d}$$

Energy produced = 50 KJ/gmethane

Density at $35^{O}C$ = 0.6346 g/L

Amount of energy produced = (719868.9) × (0.6346) × 50 =23.3×10^{6}KJ/d

REFERENCES

Arceivala, S., & Asolekar, S. R. (2007). Wastewater Treatment for Pollution Control and Reuse (3rd ed.). Academic Press.

Daud, M. K., Rizwi, H., Akram, M. F., Ali, S., Rizwan, S., Nafees, M., & Jin, Z. S. (2018). Review of Upflow Anaerobic Sludge Blanket Reactor Technology: Effect of Different Parameters and Developments for Domestic Wastewater Treatment. *Journal of Chemistry*, *2018*, 1–13. doi:10.1155/2018/1596319

Gupta, D., & Singh, S. K. (2015). Energy use and greenhouse gas emissions from wastewater treatment plants. *International Journal of Environmental Engineering*, *7*(1), 69251. doi:10.1504/IJEE.2015.069251

Harrou, F., Madakyaru, M., & Sun, Y. (2017). Improved nonlinear fault detection strategy based on the Hellinger distance metric: Plug flow reactor monitoring. *Energy and Buildings*, *143*, 149–161.

Kannan, A., & Singaram, J. (2012). Anaerobic Sequencing Batch Reactor and its influencing factors: An overview. *Indian Journal of Environmental Health*, *54*(2), 317–322.

Krishna, H. (n.d.). Review of Research on Bio-reactors used in wastewater treatment for production of bio hydrogen: future fuel. *International Journal of Science Inventions Today, 2*(4), 302-310.

Mahavi, A. H. (2008). Sequencing batch reactor: A promising technology in wastewater treatment. *Iranian Journal of Environmental Health Sciences & Engineering*, *5*(2), 79–90.

Mahvi, A. H., Mesdaghinia, A., & Karakani, F. (2004). Feasibility of continuous flow sequencing batch reactor in domestic wastewater treatment. *American Journal of Applied Sciences*, *1*(4), 348–353. doi:10.3844/ajassp.2004.348.353

Mojiri, A., Aziz, H. A., Zaman, N., Aziz, S., & Hamidi, A. (2012). A Review on Anaerobic Digestion, Bio-reactor and Nitrogen Removal from Wastewater and Landfill Leachate by Bio-reactor. *Advances in Environmental Biology*, *6*(7), 2143–2150.

Ødegaard, H. (1999). The moving bed biofilm reactor. *Water Environmental Engineering and Reuse of Water, 575314*, 205-305.

Ridgers, T. (2013). *Chemical Engineering*. Retrieved from https://personalpages.manchester.ac.uk/staff/tom.rodgers/documents/CRE_Notes.pdf

Rosen, A. (2014). *Reactor Design*. Retrieved from https://sites.tufts.edu/andrewrosen/files/2013/09/reactor_design_guide1.pdf

Stenstrom, M., & Rosso, D. (2003). *Fundamentals of Chemical Reactor Theory*. Academic Press.

Section 3
Wastewater Treatment

Chapter 3
Wastewater Treatment Operations

ABSTRACT

Sewage is treated by a variety of methods to make it suitable for its intended use, be it for spraying onto irrigation fields (for watering crops) or be it for human consumption. Sewage treatment mainly takes place in two main stages: primary and secondary treatment. In arid areas, where there is not enough water, sewage also undergoes a tertiary treatment to meet the demands of the drinking water supply. During primary treatment, the suspended solids are separated from the water and the BOD (biochemical oxygen demand) of the water is reduced, preparing it for the next stage in wastewater treatment. Secondary treatment consists of aeration and settling tank. This process removes 75-95% of the BOD. In case of trickling filter, BOD removal is up to 80%-85%. The water is then disinfected, mostly by chlorination, and released into flowing streams or oceans. Therefore, the main objective of this chapter is to provide a deeper insight into preliminary, primary, secondary, and tertiary treatment of wastewater and furthermore provide cognizance concerning design considerations of treatment units.

PRELIMINARY TREATMENT PROCESS

The purpose of preliminary treatment is to ensure a satisfactory quality of final effluent and final sludge product, removal of waste water constituents such as rags, sticks, floatables, grit and grease that may cause maintenance

DOI: 10.4018/978-1-5225-9441-3.ch003

or operational problem with the treatment operations, processes and ancillary systems (Crites & Tchobanoglous, 1998) and to protect the treatment process from malfunction associated with accumulation of screenings, debris, inorganic grit, excessive scum formation or loss of efficiency associated with grease or oil films or fat accumulations (Environmental Protection Agency Ireland, 1995) They are also referred as the physical unit operations. The unit operations used are screening for removing floating grit chambers or detritus tanks for removing grit and sand; skimming tanks for removing oils and grease; and primary settling tank or primary clarifier for removal of residual settle able suspended matter. It helps to reduce the BOD by 15% to 30% (Singh, Anurag).

Screening

In the waste water treatment generally the screening process is adopted as first unit operation. It's a device of uniform size openings using circular bars or the rectangular bars, use to retain the floating material or solids found in the influent waste water to the treatment plant. The basic principle of screening is to remove or separate the coarse floating material found in the waste water which could damage the electrical or subsequent process equipment (pumps, valves, pipe lines, impellers etc.), can inhibit or reduce the treatment process reliability & effectiveness, or contaminate waste way (Jamal, Haseeb, 2017). It consists of passing the raw sewage through different type of screens systems (Trash Rack, Manually Cleaned Bar Screen and Mechanically Cleaned Bar) with different sizes (U.S. EPA, 2000) (Fine screening, for a spacing under 10 mm, Medium screening, for spacing of 10 to 40 mm, Coarse screening, for spacing of over 40 mm).

Manual Screening

Manually cleaned bar screen (large in size, in order to reduce the frequency of screenings collection operations) at low flow rate for less floating material collecting at the screens, having bar spacing of 20 - 25 mm. Manually cleaned bar screens are used at the beginning of diverse wastewater treatment plants, also used as standby screening for servicing or repairing of the main screen or at the time of power failure, the length of the manually cleaned bar screens shouldn't be larger than the distance not covered (distance which is not faceable for the workers to clean up the bars manually) or not convenient for racked by hands (3 m or 10ft). Perforated plates attached with the screens are provided at the top for storing the rack which has been collected or stored temporarily for drainage of the excess wastewater in the rack. The influent channel and the screens should be perpendicular to each other for uniform distribution of floating material and solids throughout the flow and on the screen.

Mechanical Screening

It is the practice of taking granulated ore material and separating it into multiple grades by particle size using mechanical equipment's which are self-operated system. Mechanical screens are comes under the play, mainly for reduce the workforce and for reducing the maintenance, repairing and operating problems occurring during the screening.

Screens can be broadly classified depending upon the opening size provided in table: 2; three general types of screens are being used now a day (coarse screen, medium screen, fine screen) and may be defined as follows:

Table 1. Types of screening

Principle Types			
Chain Driven Screens	**Reciprocating Rack (Climber Screens)**	**Catenary Screens**	**Continuous Belt Screens**
Automatic rotating chain with a paddle for cleaning the screens after clogging, it classified as clock and anti-clock wise rotation on the basis of how the screens are racked depends on the upstream or downstream of the flow.	It is developed to reduce the maintenance to previous screening technologies, which used permanently submerged parts such as sprockets and bearings. - Cogwheels move the rake arm down the pin rack upon activation. - The rake arm enters the channel upstream from the screen. - At the bottom of the pin rack, the rake engages the screen. - Cogwheels walk the rake arm up the pin rack, transporting screenings for removal.	It is a screen with the front clean front return bar system, uses the two (2) heavy duty strands of steel chain which has superior strength than any other screens system, suspended with sprockets at the top shaft, as the lower end of the chain forms the "catenary" that's why this is known as catenary screens. Some proven features are. - Heavy duty - No underwater moving parts - Shafts - Bearings - Guides - Sprockets - Jam proof - None clogging trapezoidal shaped screen bars.	Use to remove the fine coarse floating material, it is a self-cleaning screens, a large no. of racks or screens are attached with the drive chain system, the structure material can be stainless steel structure consist by sturdy frame which allow the stability in small and large sizes in height and width. The structure optimum inclination varies from 65° to 85°.

Table 2. Screen openings

Screen Type	Aperture
Coarse screen	>50 mm
Medium screen	15 – 50 mm
Fine screen	3 – 15 mm
Straining	<3 mm

- **Coarse Screen**: Generally used for larger volume of discharge of wastewater, for remove the trash racks and large objects, to prevent the pumps valves, pipelines and other appurtenance from damage or clogging. Requirement and designing (manual and mechanical) of the coarse screens are depending on the characteristics of influent wastewater. Types are: manual, mechanical, chain or cable driven, reciprocating rack, catenary screens and continuous self-cleaning screens.
- **Medium Screen**: These screens are generally fitted with mechanical, the operation of the racking mechanism is periodic and intermitted, controlled by timer or head loss measurements, working of the pump is controlled by the timer. Rectangular coarse screen are generally used now a days in sewage treatment plant with an inclination of 30° to 60° to the direction of flow for the purpose of increasing the surface area of the screening process. Types are: inclined fixed (permanently fixed at position and not movable or can't remove for the cleaning), horizontal reciprocating, rotatory drum.
- **Fine Screen:** Help to reduce the suspended solids from the influent wastewater to preliminary treatment process. They can reduce 20% of suspended solids from the wastewater. Types are: tangential, rotatory drum, rotatory vertical disk, vibrating screen.
- **Straining, Very Fine and Micro Screen:** Reduce suspended solids; upgrade secondary effluent to tertiary standards. Used at very low speed or flow rate of influent wastewater with continuous backwash, it comes under the different types of filtering fabrics. The suspended solids removal is 10% to 80% with an average of 55%.

Screening Safety

As any treatment facility need some precaution, maintenance, and safety over using, the also screens in the preliminary treatment process need some safety to reduce the problems arising over functioning and to increase the efficiency and the life of the screens. Some of the common problems arise during screening are given in Table 3.

Velocity of the flow just ahead and through the screens varies with time and may reduce the efficiency of the screens and also affect the operation. The lower the velocity of the flow through the screen, the greater the amount of screening that would be removed from the wastewater. The design velocity should be of 100% removal efficiency. The satisfactory velocity through the open area for the peak flow is 0.6 to 1.2 mps and at low flow in the approach channel should not be less than 0.3 mps to avoid deposition of solids.

Head loss in the flow varies with the quantity and the nature of the screenings which is allowed to accumulate between screenings. The head loss from the screens can be calculated by flow and the effective area of the screen opening. The head loss through the screens can be calculated using theses formulas:

Table 3. Type of problems associated during screening

Screening Operational Problems		
Problems	**Cause**	**Solution**
Intolerable odour and vermin	Improper storage or prolonged storage of screening.	Provide proper storage facility; increase the frequency of removal or disposal of stored screenings.
Excessive screen clogging	Unusual amount of debris in wastewater. Low velocity through the racks. Automatic rack action not often enough.	Identify the source and provide the coarse screen. Reset the timer or install the level override.
Excessive grit accumulation	Low velocities in the channel.	Remove the irregularities in the flow. Slope the floor. Rack the channel. Flush regularly.
Jammed racking mechanism will not reset	Obstruction still in the screens. Moment setting too fine.	Remove obstruction periodically. Consult the manufacturer regarding the setting adjustment of the screen.
Screen not been racked but motor is running	Brocken chain, cable or limit switch.	Inspect the chain, racks and switches also replace them if necessary.
Rake will not operate – no visible reason	Defective controlled mechanisms.	Check control circuit, maintain, repair or replace if necessary.

Head Loss Through the Cleaned Flat Bars

$$h = 0.0729\left(V^2 - v^2\right) \tag{1}$$

where the

h = head loss in m
V = velocity through the screen in m/s
v = velocity before the screen in m/s.

Head Loss Through the Coarse Screen

$$h_l = \frac{1}{C}\left(\frac{V^2 - v^2}{2g}\right) \tag{2}$$

where,

h_l = head loss in m
C = an empirical discharge coefficient to account for turbulence and eddy losses, typically 0.7 for a clean screen and 0.6 for a clogged screen
V = velocity of flow through the openings of the bar screens in m/s
v = approach velocity in upstream channel in m/s
g = acceleration due to gravity, 9.81 m/s^2

Head Loss Through a Bar Rack

$$h = \beta\left(\frac{W}{b}\right)^{\frac{4}{3}} h_v \sin\theta \tag{3}$$

h = head loss, m

β = bar shape factor (2.42 for sharp edge rectangular bar, 1.83 for rectangular bar with semicircle upstream, 1.79 for circular bar and 1.67 for rectangular bar with both u/s and d/s face as semi-circular).
W = maximum width of bar u/s of flow, m
b = minimum clear spacing between bars, m
h_v = velocity head of flow approaching rack, m = $v^2/2g$
θ = angle of inclination of rack with horizontal

The Head Loss Through the Fine Screen

$$h = \left(\frac{1}{\left(2g.C_d\right)}\right)\left(\frac{Q}{A}\right)^2 \qquad (4)$$

where,

g = gravity acceleration (m/sec^2)
C_d = coefficient of discharge = 0.6 for clean rack
Q = discharge through screen (m^3/sec)
A = effective open submerged area (m^2)

Screening Removal Computations

Examples 1

Design a bar screen chamber for average sewage flow 20 MLD, minimum sewage flow of 12 MLD and maximum flow of 30 MLD.
Sol.

1. Average flow = 20 MLD = 0.231 m^3/s

Maximum Flow = 30 MLD = 0.347 m^3/s

Minimum flow = 12 MLD = 0.139 m^3/s

2. Assume manual cleaning and angle of inclination of bars with horizontal as 30°. Assume size of bars 9 mm x 50 mm, 9 mm facing the flow. A clear spacing of 30 mm between the bars is provided.

3. Assume velocity of flow normal to screen as 0.3 m/sec at average flow.
4. Net submerged area of the screen opening required

$$= \frac{0.231 m^3 / \text{Sec}}{0.3 m / \text{sec}} = 0.77 m^2$$

Assume velocity of flow normal to the screen as 0.75 m/sec at maximum flow, hence net submerged area of screen opening

$$= \frac{0.347 m^3 / \text{Sec}}{0.75 m / \text{sec}} = 0.46 m^2$$

Provide net submerged area = 0.77 m^2

5. Gross submerged area of the screen When 'n' numbers of bars are used the ratio of opening to the gross width will be;

$[(n+1)30] / [(n+1)30 + 9 \text{ x } n] \approx 0.77$

(For 20 to 30 numbers of bars)
Therefore gross submerged area of the screen = 0.77 / 0.77 = 1 m^2

6. The submerged vertical cross sectional area of the screen = 1.0 x Sin 30 = 0.5 m^2

This is equal to c/s area of screen chamber, therefore velocity of flow in screen chamber = 0.231 / 0.5 = 0.462 m/sec
This velocity is greater than the self-cleansing velocity of 0.42 m/sec

7. Provide 30 numbers of bars. The gross width of the screen chamber will be:

= 30 x 0.009 + 31 x 0.03 = 1.2 m

Therefore, liquid depth at average flow = 0.5 / 1.2 = 0.416 m
Provide free board of 0.3 m
Hence, total depths of the screen = 0.416 + 0.3 = 0.716 m, say 0.75 m
Thus, the size of the channel = 1.2 m (width) x 0.75 m (depth)

8. Calculation for bed slope:

R = A/P = (0.416 x 1.2) / (2 x 0.416 + 1.2)

R= 0.246 m

Now,

$V = (1/n)\ R^{2/3}\ S^{1/2}$ (5)

$S^{1/2} = V.n / R^{2/3} = 0.462 \times 0.013 / (0.246)^{2/3}$

$S^{1/2} = 0.0153$

Therefore bed slope is nearly 1 in 4272 m

9. Head loss through the screen, h, when screen is not clogged

$$h = 2.42\left(\frac{9}{30}\right)^{\frac{4}{3}}\left[\frac{\left(0.462\right)^{2}}{2\times 9.81}\right]\sin 30$$

$= 2.65 \times 10^{-3}$ m

= 0.00265 m

= 2.65 mm

For half clogged screen, the head loss can be worked out using opening width as half

Thus, b = 30/2 = 15 mm

And $h = 6.67 \times 10^{-3}$ m

= 6.67 mm < 150 mm

However, provide 150 mm drop of after screen.

If this head loss is very excessive, this can be reduced by providing bars with rounded edges at upstream, or by reducing width of bars to 6 to 8 mm, or by slight reduction in velocity.

Except for the change in shape of bars in other cases the channel dimensions will change. For minimum flow and maximum flow, the depth of flow can be worked out using Manning's formula using known discharge, and check for velocity under both these cases, as self-cleansing and non-scouring, respectively, and also depth of flow at maximum discharge.

Example 2

Design the screening chamber for preliminary treatment process of wastewater treatment plant having 50 MLD discharge.

Sol. Given discharge is 50 MLD

1. Q = 50 MLD = 0.578 m^3/s

Assume, velocity of through the screen is 0.8 m/s maximum at the peak flow.

Therefore Area of screen require

$$Q = A \times V \tag{6}$$

$$A = \frac{0.578}{0.8} = 0.7233m^2$$

Assume the depth of the screen (D) is 1 m

Therefore width of the screen (B) = 0.7233/1 = 0.7233 m

2. Let the size of the screens are design for medium screen with the opening of 40 mm.

Therefore the no. of openings in the screen chamber = 0.7233 m/40 mm

= 18.0825 say 18

Hence the number of bars in the screen are = 18+1 = 19

Gross width of the screen (B')

B' $= 0.04 \times 18 + 19 \times 0.01 = 0.91$

Gross area of the screen (A') $= 0.91 \times 1 = 0.91m^2$

3. For increasing the screening area of the screen,

Assume that the screen bar are placed at an angle of 45°
Therefor the new gross area of the screen needed (A")

$A'' = 0.91 \times \sqrt{2} = 1.28m^2$

4. The velocity through the screen = 0.8 m/s

And the velocity above the screen = 0.578/1.28 = 0.45 m/s

Head loss through the cleaned screen $= 0.0729\left(V^2 - v^2\right)$

Figure 1. The screen is placed at an angle 45°

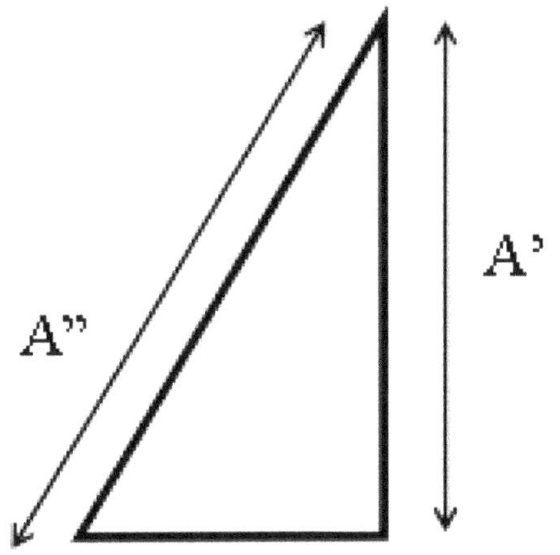

$$= 0.0729\left(0.8^2 - 0.45^2\right)$$

$$= 0.0318m = 3.18cm$$

5. When the screens are half clogged

The velocity through the screen (v) $= 0.8 \times 2 = 1.6 \text{ m/s}$

Head loss $= 0.0729\left(V^2 - v^2\right) = 0.0729\left(1.6^2 - 0.45^2\right) = 0.171m = 17.1cm$

Since the head loss is very high therefore the cleaning of the screen frequently or periodically is very necessary to keep the head loss in the allowable range.

Grit Removal

Grit chamber is the second unit operation used in preliminary treatment of wastewater and it is intended to remove suspended inorganic particles such as sandy and gritty matter from the wastewater. This is usually limited to municipal wastewater and generally not required for industrial effluent treatment plant, except some industrial wastewaters which may have grit. The grit chamber is used to remove grit, consisting of sand, gravel, cinder, gritty matter, egg shell, bones, nonputrecible material (general solid waste, example: glass, plastic, rubber, plasterboard, ceramics, bricks, concrete etc.) and other heavy solids materials that have specific gravity (about 2.65 and of 0.21 mm diameter) much higher than those of the organic solids in wastewater. Grit chambers are provided to protect moving mechanical equipment from abrasion and abnormal wear; avoid deposition in pipelines, channels, and conduits; and to reduce frequency of digester cleaning. Separate removal of suspended inorganic solids in grit chamber and suspended organic solids in primary sedimentation tank is necessary due to different nature and mode of disposal of these solids. Grit can be disposed off after washing, to remove higher size organic matter settled along with grit particles; whereas, the suspended solids settled in primary sedimentation tank, being organic matter, requires further treatment before disposal.

Generally there are two types of grit removal chambers:

- **Horizontal Flow:** A long narrow channel is used in this type of grit chamber. The wastewater moves through this channel in more or less plug flow condition with minimal mixing to support settling of the particles. Higher length to width ratio of the channel is used to minimize mixing. For this purpose a minimum allowance of approximately twice the maximum depth or 20 to 50% of the theoretical length of the channel should be given for inlet and outlet zones. The width of this channel is kept between 1 and 1.5 m and the depth of flow is normally kept shallow. A free board of minimum 0.3 m and grit space of about 0.25 m is provided. For large sewage treatment plant, two or more number of grit chambers are generally provided in parallel. The detention time of 30 to 60 seconds is recommended for the grit chamber.
- **Rectangular or Square Configuration:** This is an oldest type of grit chamber with the rectangular horizontal flow, velocity control type. Used to maintain the velocity close to 0.3 m/s for providing the sufficient time for grit particles to settle down to the bottom of the channel. The lightest particle of grit will reach the bottom of the chamber before reaching to its ends. The square horizontal flow grit chamber are being used by the treatment plant from 60 years. Influent from the screening chamber is introduced over the cross section of the chamber by using the vanes or gates, and the distributed wastewater flows in straight lines across the tank and overflows a weir in a free discharge. For the square grit chamber it is advisable to use the two grit camber for the grit removal, and design on the basis of over flow rate that are depends on the particle size and temperature of the wastewater, they are normally designed to remove the 95 percent of the 0.15mm diameter particle at peak flow.

Gravity/Velocity-Controlled Grit Removal

With variation in sewage flow received at the treatment plant, it is important that the velocity of the wastewater in the grit chamber should be maintained nearly constant. Otherwise, when the flow is lower, deposition of not only inorganic solids but also organic solids will occur in grit chamber due to lowering of velocity. With flow higher than average, when the velocity will exceed the critical velocity, scouring of already deposited grit particles will occur leading to failure of performance. Hence for proper functioning, the velocity should not be allowed to change in spite of the change in flow in the grit chamber. This can be achieved by the provision of the proportional weir or Par shall flume at the outlet end of the grit chamber. The shape of the opening between the plates of a proportional weir is made in such a way that the discharge is directly proportional to liquid depth in the grit chamber. As a result, the velocity of water in the chamber will remain constant for all flow conditions.

The discharge through proportional weir can be given by the following equation:

$$Q = C.b\sqrt{2ag}\left[H - \frac{a}{3}\right] \tag{7}$$

where, Q = Discharge, m^3/sec^1

C = constant, 0.61 for symmetrical sharp edged weir
a = 25 to 35 mm
b = base width of the weir
H = Height of water above the crest of weir

The equation of the curve forming the edge of the weir is given by the following formula:

$$x = \frac{b}{2}\left[1 - \frac{2}{\pi}\tan^{-1}\sqrt{\left(\frac{y}{a}\right) - 1}\right] \tag{8}$$

The sharp edges generated by the curve at the bottom are curtailed on both the side, because such small opening will not contribute for flow due to deposition of solids. These edges are curtailed from the side wall at a distance of minimum 75 mm and height of the vertical edge ‘a’ is in the range of 25 to 35 mm. To compensate this loss of area the edge of the weir is lowered by a/3 than the theoretical level.

$$Q = Cd\ (2g)^{1/2}\ L\ H^{3/2} \tag{9}$$

For normal sharp crested weir, where as in proportional weir $Q \alpha H$ instead of $H^{3/2}$

Figure 2. Schematic diagram of the front view of proportional weir
(Bengtson, 2019)

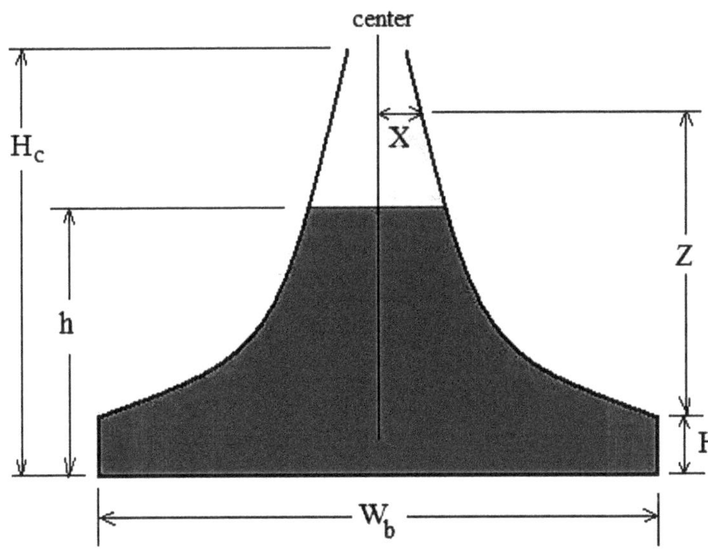

Alternatively, Par shall flume can be placed at the end of the grit chamber. The design details for Par shall flume to meet different discharges are provided in the CPHEEO manual (1993). With appropriate arrangement this will also facilitate recording of the discharge received at the sewage treatment plant.

Aeration

Aerated grit chamber: Normally design to remove the particle equal and greater than 0.21 mm with 2 to 5 minutes of detention time, the depth of the hopper is about 0.9 m with steeply sloping slide located at the one side of the chamber, air diffusers are located at about 0.45 to 0.6 m above the plane of the bottom. Airflow is produced by a blower and is introduced into the Aerated Grit Chamber via a tube which is located near the bottom of the chamber, circular or toroidal flow pattern are created in the wastewater. The continuous increasing flow deflects off an energy-recovery baffle at the liquid surface. This flow pattern causes the grit to settle to the bottom of the chamber while keeping lighter organic material in suspension used in further process. Either a recessed-impeller grit pump or, more commonly, an air-lift pump is used to remove the grit slurry settled and send it on for dewatering. Benefits of the aerated grit chamber is; maintenance is significantly reduced, consistent grit removal efficiency over a wide flow range, possible septic conditions of the plant influent may be alleviated through pre-preparation in the grit chamber, performance of downstream units may be improved by using pre-aeration to reduce septic conditions in incoming wastewater, aerated grit chambers are versatile, allowing for chemical addition, mixing, pre-aeration and flocculation. Almost 100 percent efficiency can be obtained with proper modification. Grit containing organic material can cause odour nuisance and attract insects. Using and releasing of significance amount of Volatile Organic Compound (VOC's) by the air agitation in the aerated grit chamber can be a health risk to the treatment plant operators, covers may be required or non-aerated type grit chamber can be used for prevention. Aerated grit chambers also require more power than other grit removal processes, and maintenance and control of the aeration system requires additional labor.

Design considerations for aerated grit chambers include the following (WEF 1998):

- Air rates typically range from 0.3 to 0.7 m^3/m/min (3 to 8 ft3/ft/min) of tank length
- A typical minimum hydraulic detention time at maximum instantaneous flow is two minutes.
- Typical length-to-width ratio is 2.5:1 to 5:1.
- Tank inlet and outlet are positioned so the flow is perpendicular to the spiral roll pattern.
- Baffles are used to dissipate energy and minimize short circuiting.

Cyclone Degritter

Vortex type: it is the device which used vortex flow pattern to remove the grit particle from the wastewater. It is basically of two types; first type, the wastewater enters and exists tangentially. Rotating turbines are used to maintain the flow velocity, and the adjustable pitch blades promote the separation of organic material from the grit, propellers produce the toroidal flow path for grit. Then this grit settles down by the action of gravity into the hopper of the basin in one resolution, then the grit removed by the grit pump. The pumped grit then can be discharge to a hydro clone for the removal of remaining organic material. Special arrangements for the flow splitting are required for the two grit chambers.

Second type, vortex is generated with the help of flow which is entering tangentially from the top of the chamber; effluent exits the chamber from centre of the top of the unit from a rotating cylinder (eye of the fluid). By the action of centrifugal force in the flow the particles with densities greater than water are catches on the wall and then settle down with the help of gravitational force, the grit chamber used to handle peak flow rate up to 0.3 m^3/s per unit, grit from the chamber then removed by using the cleated belt conveyor. This type of design requires deep basement or lifting arrangement for installation.

Grit Removal Calculations

Example 3

Design a grit chamber for population 50000 with water consumption of 135lpcd.
Sol.

1. Average quantity of sewage, considering sewage generation 80% of water supply, is = 135 x 50000 x 0.8 = 5400 m^3/day = 0.0625 m^3/sec
2. Maximum flow = 2.5 x average flow = 0.0625 x 2.5 = 0.156 m^3/sec

Keeping the horizontal velocity as 0.2 m/sec (<0.228 m/sec) and detention time period as one minute.

3. Length of the grit chamber = velocity x detention time = 0.2 x 60 = 12.0 m
4. Volume of the grit chamber = Discharge x detention time = 0.156 x 60 = 9.36 m^3
5. Cross section area of flow 'A' = Volume / Length = 9.36/12 = 0.777 m^2
6. Provide width of the chamber = 1.0 m, hence depth = 0.777 m
7. Provide 25% additional length to accommodate inlet and outlet zones.

Hence, the length of the grit chamber = 12 x 1.

25 = 15.0 m

8. Provide 0.3 m free board and 0.25 m grit accumulation zone depth,

Hence total depth = 0.777 + 0.3 + 0.25 = 1.33 m and width = 1.0 m

Example 4

Design a horizontal flow grit chamber with rectangular cross section for treating maximum sewage flow of 10 MLD at maximum temperature of 34° C during summer and minimum temperature of 15° C in winter.

Sol. The settling velocity of the grit particle will be less at lower temperature, i.e., 15° C.

1. At this temperature kinematic viscosity = 1.14 x 10-2 cm²/sec in first trial assume Reynolds number 'R' less than or equal to 0.3.

$$V_s = \frac{g}{18}\left[\frac{\rho_p - \rho_w}{v}\right] \times D^2 \qquad (10)$$

$$V_s = \frac{981}{18}\left[\frac{2.65-1}{1.14\times 10^{-2}}\right]\times 0.02^2 = 3.15cm / \sec$$

Reynolds Number R = v.D/ν = 3.15 x 0.02 / 1.14 x 10-2 = 5.53 > 0.3

2. Therefore, Vs is not equal to 3.15 cm/sec because the equation for Vs is valid only for R < 0.3. Using Vs = 3.15 cm/sec, calculate R and CD and then again Vs till it converges.
3. Subsequent Trial

Vs = 2.4 cm/sec

Table 4. Design parameters

1. Length of the grit chamber, with 25% additional length	15.0 m
2. Volume of the grit chamber	9.36 m³
3. Width of the chamber	0.777 m
4. Total depth of the grit chamber	1.0 m

R = 2.4 x 0.02/(1.14 x 10-2) = 4.21

$$C_D = \frac{24}{4.21} + \frac{3}{\sqrt{4.21}} + 0.34$$

C_D= 7.50

4. For equation

$$V_s = \sqrt{\left[\frac{4}{3}\frac{981}{7.50}\left(2.65 - 1\right)0.02\right]}$$

Vs = 2.4 cm/sec Hence O.K. (2074 m/d)

5. Now for β = 0.06, f = 0.03, and D = 0.02 cm

$$V_s = \sqrt{\left[\frac{8\beta}{f} g\left(\rho_p - \rho_w\right) D\right]} \quad (11)$$

$$V_s = \sqrt{\left[\frac{8 \times 0.06}{0.03} 981\left(2.065 - 1\right)0.02\right]}$$

V_S= 22.76 cm/sec

6. Now Q = 10 MLD = 0.116 m^3/sec

Therefore,

C/S Area A = Q/V = 0.116/0.227 = 0.51 m^2

7. If width of 1 m is provided, the depth required = 0.51 m

Provide total depth = 0.51 + 0.3 (free board) + 0.25 (space for grit accumulation)

= 1.06 Say 1.1 m

8. Now Vo/Vc = H/L = 2.4/22.7

Therefore

theoretical length L = 22.7 x 0.51 / 2.4 = 4.824 m

9. Provide 2 m extra length for inlet and outlet

Therefore

total length = 2 + 4.824 = 6.824 m say 6.9 m

10. Total working volume = 0.51 x 6.9 x 1 = 3.52 m^3

Hence,

Overall detention time = 3.52 / 0. 116 = 30.34 sec (within 30 to 60 seconds)

Preaeration

The benefits from the preaeration is depends on the characteristics of the wastewater, design of preliminary treatment units, amount and the methods of application of aeration. The BOD removal efficiency in the preliminary treatment with the help of preaeration is depends the amount of aeration,

Table 5. Design parameters

1. Length of the grit chamber	6.9 m
2. Working volume of the grit chamber	3.52 m^3
3. Width of the chamber	1 m
4. Total depth of the grit chamber	0.51 m
5. Over all detention time	30 – 60 seconds

detention time and the method adopted for the aeration. The preaeration in the aerated grit chamber is has the significance effect, it will remove the organic material from the surface of the grit particles.

Aeration tanks are characterized by the introduction of air bubbles into the water. Depending on their size, the bubbles can be: fine bubbles (with diameter between 1-1.5 mm) ; medium bubbles (diameter 1.5-3 mm); large bubbles (diameter over 3 mm). Fine bubbles are obtained by porous diffusers or air distribution through membranes with very fine holes. Medium bubbles

Figure 3. Schematic diagram of pre-aerated Grit Chamber

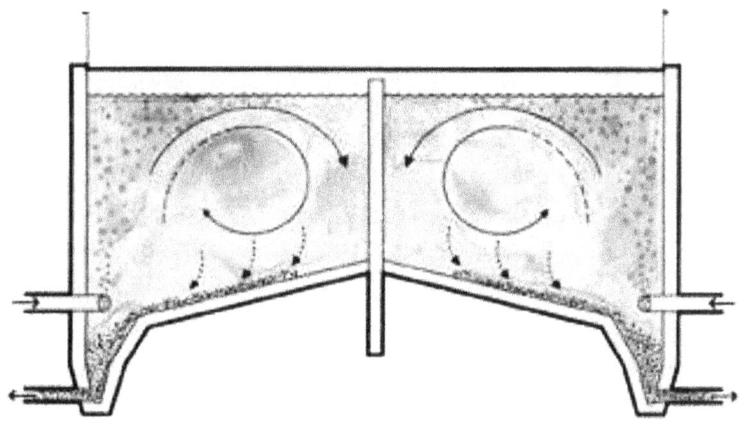

Figure 4. (a) and (b) Fine bubbles aearators

(a)

(b)

resulting from the air distribution tubes provided with orifices whose diameter is between 1-5 mm and is spaced less than 5 mm from each other. Large bubbles are the result of air distribution through pipes or plates drilled with holes opening over 5 m.

If mechanical adhesion achieves strong water agitation in the tank, it produces intense mixing between water, mud and air. Considering the way in which the air suction is made, mechanical aeration systems are shown in Figures 5-9.

Figure 5. Down-draft, induced-aspiration aerator

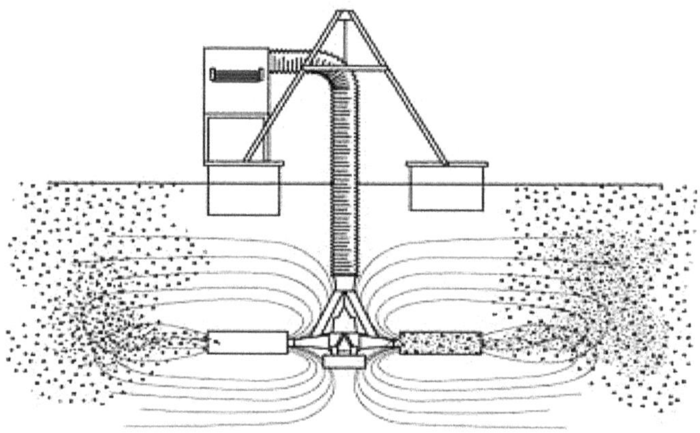

Figure 6. Mechanical rotor aerators

Figure 7. Mechanical aerators with blades or brushes

Mechanical aeration tanks with aspiration are rarely used. These aerators are actually vertical tubes through which the air is sucked with water through pipes, known as the Ventury effect and water + air mixture is pushed to the bottom. In the case of mixed aeration methods, the two schemes mentioned above are used which achieves the dispersion of air into the water table (pneumatic) and forcing convection by mechanical system (system recommended in waters heavily organic loaded). Regardless the aeration system, there must be taken into consideration the amount of air circulated and how the transfer of oxygen from air to water is achieved. Using oxygen from the air for biological wastewater treatment depends on the variation of physical quantities (pressure, temperature, etc.) and the characteristics of wastewater (biodegradability of organic materials, their amount in the water, the oxygen content initially).

Chemical Addition

There are numerous treatment methods and techniques used for the separation of solids from the wastewater in influent and effluent water. From which one of the effective method is coagulation and flocculation, different types of chemicals are being used to separating and disinfecting the solids from the wastewater, lime softening, sludge thickening, and solids dewatering and removal, and also heavy metals removal, oil/water emulsions, paint detoxification, odor control, and defoaming. In the influent from any effluent sources, very fine suspended particles arc also present in very high quantity specially in sewage water which hard to removed by settle using mechanical methods, may be settled by making the flocculated particles which are heavier the water, using the chemicals. Some common chemical coagulants used in the wastewater treatment plants are: Alum (aluminum sulfate), $Al_2(SO_4)_3$. Still, the most common coagulant in the United States, it is often used in conjunction with cationic polymers, PAC, Poly Aluminum Chloride, Polyaluminum chloride, Al(OH)x(C1)y. This is efficient in some waters, requiring less pH adjustment and producing less sludge.(simply its called PAC), Ferric chloride, $FeCl_3$. This may be more effective than alum in some applications, Ferric sulfate, $Fe_2(SO_4)_3$. It is effective in some waters and more economical in some locations, Ferrous Sulfate, FeSO4. It costs bit more compare with ferric chloride but result would be better than that, Sodium Aluminate, $Na_2Al_2O_4$, Cationic polymers can be used alone as the primary coagulant or in conjunction with aluminum or iron coagulants.

Coagulants neutralize the negative electrical charge on particles (colloidal fine particles), which destabilizes the forces keeping colloids isolated. These coagulants contain positively charged molecules that, when added to the water and mixed thoroughly, accomplish this charge neutralization. Inorganic coagulants, organic coagulants or a combination of both are typically used to treat wastewater for removal of suspended fine particles. Charge-neutralized solids can be further agglomerated by using flocculants. Flocculants are sort of a "high-tech rope" tying particles together, thereby increasing particle size and density. Flocculants come in various forms on their charges, charge densities, molecular weights etc. ChemTreat's portfolio includes Cationic Flocculants; Mostly based on copolymers of AETAC (N, N-Dimethylaminoethyl Acrylate Methyl Chloride Quaternary) or METAC (N, N-Dimethylaminoethyl Methacrylate Methyl Chloride Quaternary) and acrylamide. These products can perform a dual function by both coagulating with their positive ionic charge and flocculating with their high molecular weight. Anionic Flocculants; Mostly based on copolymers of acrylamide and acrylic acid, anionic flocculants possess a negative ionic charge and work by binding with residual cationic charges on coagulants adsorbed to coagulated colloids.

However there are some merits and demerits of chemical treatment of wastewater as simple, economically feasible, chemical method comparatively mature, preliminary treatment aided with coagulation give the better results in BOD and suspended particles, also required the less space for the treatment of the same, also remove the phosphate from the influent helps to control the eutrophication. Although these are the cause of high amount of sludge production in the plant as high amount of chemicals are used to maintain the pH, makes some biological treatment processes difficult, inhibits the growth of the microorganisms for further treatment processes thus creating the difficulties in sludge digestion, required skilled labour and supervisor and makes the treatment costlier.

Equalization

Equalization is the operation aimed at eradicating or reducing the difficulties associated with fluctuations in wastewater flow and characteristics. They consist of holding the influent for some proposed time in the continuous mix basin, for the homogenous mixture, this arrangement is, of course, necessary, when the wastewater comes from the industries, having different quantities and characteristics throughout the day. Various benefits are ascribed by different investigators to the use of flow equalization in wastewater treatment systems. Some of the most important benefits are listed as follows; Equalization improves sedimentation efficiency by improving hydraulic detention time, The efficiency of a biological process can be increased because of uniform flow characteristics and minimization of the impact of shock loads and toxins during operation, Manual and automated control of flow-rate-dependent operations, such as chemical feeding, disinfection, and sludge pumping, are simplified,Treatability of the wastewater is improved and some BOD reduction and odor removal is provided if aeration is used for mixing in the equalization basin, A point of return for recycling concentrated waste streams is provided, thereby mitigating shock loads to primary settlers or aeration basin (Goel, Flora, & Chen, 2010).

The objective of equalization is to minimize or control fluctuations in wastewater characteristics in order to provide optimum conditions for subsequent treatment processes, to provide adequate dampening of organic fluctuations in order to prevent shock loading of biological systems, to provide adequate pH control or to minimize the chemical requirements necessary for neutralization, to minimize flow surges to physicalchemical treatment systems and permit chemical feed rates compatible with feeding equipment, to provide continuous feed to biological systems over periods when the manufacturing plant is not operating, to provide capacity for controlled discharge of wastes to municipal systems in order to distribute waste loads more evenly, to prevent high concentrations of toxic materials from entering the biological treatment plant. Equalization basins in a treatment system can be located in-line or side-line. In-line and side-line equalization systems are both effective ways to equalize flow volumes, but in-line equalization is more effective for leveling out the variations in influent concentration because the entire flow is blended with the entire contents of the holding tank.

Air is sometimes injected into these basins to provide: better mixing; chemical oxidation of reduced compounds; to achieve some degree of biological oxidation; and agitation to prevent suspended solids from settling. Computation of the volume of an equalization basin is the key design requirement and is based on inflow variation over time. There are two methods used to compute equalization volume. One procedure is based on the characteristic regular flow pattern, the other is based on the mass loading pattern of a particular component. The first method relies on computing the equalization volume based on the excess daily average flow storage. The required volume is determined graphically by constructing a hydrograph. The function of the basin is to store flows in excess of the average daily flow and to divert this flow during times when the inflow is less than the average daily flow. The second method computes the volume based on mass loading variations within an acceptable range.

- The first step is to draw a cumulative volume curve, based on the wastewater flow; the volume is also calculated from the flow. The cumulative volume is achieved by adding the volume at the start of a pre-selected time period to the volume in the next time period. The resulting volume is then added to the volume in the subsequent time period. This process is continued until a cycle of low-flow and high-flow is completed (typically 24 h).

Figure 8. Schematic Design of (A) In line and (B) Side line system

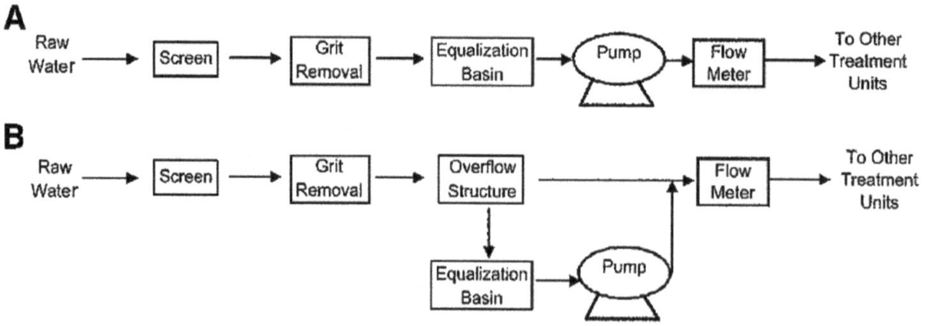

- The second step is to determine the required equalization volume by drawing a line parallel to the average flow rate and tangent to the cumulative influent flow diagram. The equalization volume is calculated by the vertical distance from the point of tangency to the straight line. There could be several points of tangency on the cumulative influent flow curve. However, care should be taken in selecting the points to be taken into consideration for equalization volume calculations.

Example 5

Determine the volume required for the equalization tank for the following flow rate given in table 4.

Sol.

1. Average pumping = 193.3 m^3/24 h = 8.054 m^3/h, hence in three hours pumping volume of wastewater pumped = 24.1625 m^3
2. From the table after calculating maximum cumulative deficit and surplus, the volume of equalization basin required = 42.1875 + 1.86 = 44.047 m^3
3. Provide 20% extra volume, hence volume of the tank = 53 m^3
4. Provide mixer of capacity 0.004 KW/m^3
5. Therefore, power required for mixer = 53 x 4 = 212 W
6. Hence provide mixer of about 250 W to impart mixing in the equalization basin.
7. Provide depth of the basin = 3.5 m, hence area required = 15.14 m^2
8. Provide suitable square or circular tank.

Skimming Tank

Fats, waxes, fatty acid, soap, minerals and vegetable oil present in waste water are collectively called as oil and grease. As oil and grease are lighter than water they are normally separated by natural flotation. The bubbles of air are passed on the bottom of the tank and floating matter rises and remain on the surface of waste water which can be separated easily. It is a chamber so arranged that floating matter rises and remains on the surface of wastewater until removed, while liquid flows out continuously through deep outlets or under partition or deep scum board. This may be accomplished in a separate tank or combined with primary sedimentation. In conventional

Table 6. Variation in the flow rate of the wastewater

Time Period	Volume of Wastewater, m^3	Cumulative Volume, m^3	Cumulative Pumping, m^3	Cumulative Surplus, m^3	Cumulative Deficit, m^3
7 – 10	22.3	22.3	24.162		**1.86**
10 – 13	43.2	65.5	48.325	17.175	
13 – 16	16.8	82.3	72.49	9.81	
16 – 19	41.1	123.4	96.65	26.75	
19 – 22	39.6	163	120.812	**42.187**	
22 – 01	11.1	174.1	144.975	29.125	
01 – 04	11.1	185.2	169.137	16.063	
04 – 07	8.1	**193.3**	193.3	0	

sewage treatment plants, separate skimming tanks are not used, unless specifically required, and this is achieved by providing baffle ahead of the effluent were in the primary sedimentation tank. Skimming tanks are used to remove lighter, floating substances, including oil, grease, soap, pieces of cork and wood, vegetable debris, and fruit skins. The tank can be rectangular or circular, designed for the detention period of 1 to 15 minutes. Typical detention time of about 5 min is adopted in design. The submerged outlet is located opposite the inlet and at a lower elevation to assist in flotation and remove any solids that may settle.

The chamber is a long trough-shaped structure divided up into two or three lateral compartments by vertical baffle walls having slots for a short distance below the sewage surface and permitting oil and grease to escape into stilling compartments.

The rise of the floating matter is brought about the blowing air into the sewage from diffusers placed in the bottom. Sewage enters the tank from one end, flows longitudinally and leaves out through a narrow inclined duct. A theoretical detention period of 3 minutes is enough. The floating matter can be either hand or mechanically removed.

Grease traps are in reality small skimming tanks designed with submerged inlet and the bottom outlet Figure 9. The traps must have sufficient capacity to permit the sewage to cool and grease to separate. Frequent cleaning through removable covers is essential for satisfactory operation. Grease traps are commonly employed in case of industries, garages, hotels, and hospitals. Oil skimming devices come in a variety of types with varying features. Oil Skimmers can be used for tanks, pits, ponds, lagoons, basins, wash systems,

Figure 9. Schematic design of skimming tank

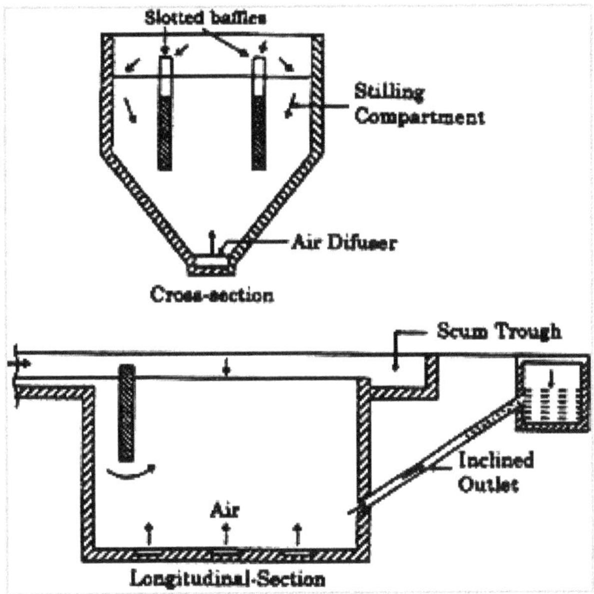

Figure 10. Schematic design of grease trap

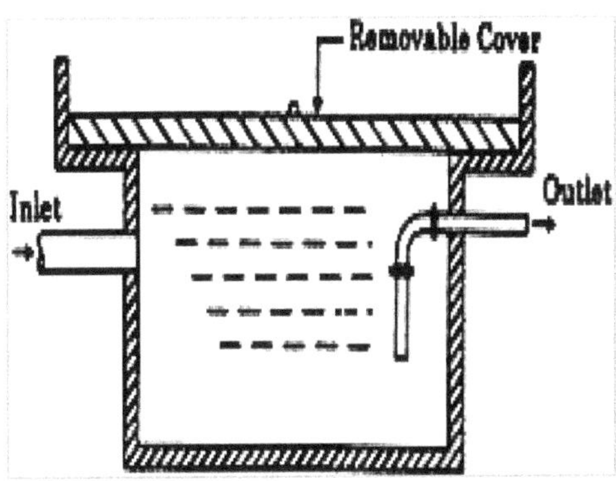

API separators, clarifiers, sumps, and many more purposes. The Sparrow Hawk 2 and the Barracuda 3 are two unique oil skimmers.

The Sparrow Hawk 2 Belt Mop Oil Skimmer is both quick and efficient when it comes to the removal of tramp oils from coolant tanks, wash water sumps or any small area where hydrocarbons accumulate. This unit is designed for both vertical and horizontal recovery of floating oils and is fairly simple to operate. A single, continuous loop of oil-attracting "mop" drops from the unit, travels across the surface of the contaminated area, and then returns to the unit where the oil is squeezed from the mop by squeegee rollers inside.

The Barracuda 3 Rope Mop Oil Skimming System is designed for small industrial applications such as sumps, basins or any in-plant location where tramp oil collects. It features two 3" diameters, electricity powered squeegee rollers that are capable of pulling up to fifty feet of rope mop. It can be used as either a belt mop or rope mop in both horizontal and vertical applications. The operation is simple with one twist, the oil attracting "mop" falls from the mop engine, journeys across the outward part of the area to be skimmed, then is pulled back to the mop engine where the oil is being pressed from the mop.

Preliminary Treatment Sampling and Testing

Sampling Methods

If an environmental domain was completely homogeneous, a single sample would adequately represent it. However, we seldom come across such a situation, as the environment is highly heterogeneous. A static system is one which does not change much with time. It must be sampled so that the sample reflects all the inhomogeneity of the system. If a field is to be tested for a long-lived pesticide in the soil, that could be considered to be a relatively static system. A dynamic system is one whose content changes with time. Most regions which we wish to characterize by taking samples are dynamic to some extent and show both spatial and temporal variation. When a river or a waste effluent stream is to be characterized, its concentration will probably change over a period of minutes, days, or hours. Systematic, Random, Judgmental (no statistical), Stratified, Haphazard, are the common sampling methods.

Types of Samples;

- **Grab Sample:** A grab sample is a discrete sample which is collected at a specific location at a certain point in time. If the environmental medium varies spatially or temporally, then a single grab sample is not representative and more samples need to be collected.
- **Composite Sample:** A composite sample is made by thoroughly mixing several grab samples. The whole composite may be measured or random samples from the composites may be withdrawn and measured.

The characteristics of sewage can be classified under following three heads: Physical Characteristics, Chemical Characteristics, and Biological Characteristics.

Primary Treatment

The purpose of preliminary treatment is to protect the operation of the wastewater treatment plant. This is achieved by removing from the wastewater any constituents which can clog or damage pumps, or interfere with subsequent treatment processes. Preliminary treatment devices are, therefore, designed to: (1) Remove or to reduce in size the large, entrained, suspended or floating solids. These solids consist of pieces of wood, cloth, paper, plastics, garbage, etc. together with some fecal matter. (2) Remove heavy inorganic solids such as sand and gravel as well as metal or glass. These objects are called grit. (3) Remove excessive amounts of oils or greases. A number of devices or types of equipment are used to obtain these objectives.

Process Description

Both physical and chemical methods are used in this treatment level, Chemical unit processes: - Chemical unit processes are always used with physical operations and may also be used with biological treatment processes. Chemical processes use the addition of chemicals to the wastewater to bring about changes in its quality. Example: pH control, coagulation, chemical precipitation and oxidation. pH Control: - To adjust the pH in the treatment process to make wastewater pH neutral. For acidic wastes (low pH): NaOH, Na_2CO_3, $CaCO_3$ or $Ca(OH)_2$. For alkali wastes (high pH): H_2SO_4, HCl. Chemical coagulation and Flocculation: - Coagulation refers to collecting the minute solid particles dispersed in a liquid into a larger mass. Chemical coagulants like $Al_2(SO_4)_3$ {also called alum} or $Fe_2(SO_4)_3$ are added to wastewater to improve the attraction among fine particles so that they come together and form larger particles called flocs. A chemical flocculent (usually a polyelectrolyte) enhances the flocculation process by bringing together particles to form larger flocs, which settle out more quickly. Flocculation is aided by gentle mixing which causes the particles to collide.

Removal of a portion of the suspended solids and organic matter from the wastewater, after grit removal in grit chamber, the wastewater containing mainly lightweight organic matter is settled in the primary sedimentation tank (PST). The objective of primary treatment is the removal of settleable organic and inorganic solids by sedimentation, and the removal of materials that will float (scum) by skimming. Approximately 25 to 50% of the incoming biochemical oxygen demand (BOD_5), 50 to 70% of the total suspended solids (SS), and 65% of the oil and grease are removed during primary treatment. Some organic nitrogen, organic phosphorus, and heavy metals associated with solids are also removed during primary sedimentation but colloidal and dissolved constituents are not affected. The effluent from primary sedimentation units is referred to as primary effluent. In many industrialized countries, primary treatment is the minimum level of pre-application treatment required for wastewater irrigation. It may be considered sufficient treatment if the wastewater is used to irrigate crops that are not consumed by humans or to irrigate orchards, vineyards, and some processed food crops. However, to prevent potential nuisance conditions in storage or flow-equalizing reservoirs, some form of secondary treatment is normally required in these countries, even in the case of non-food crop irrigation. It may be possible to use at least a portion of primary effluent for irrigation if off-line storage is provided (M.B. Pescod, 1992).

Overview of Primary Treatment

Primary sedimentation tanks or clarifiers may be round or rectangular basins, typically 3 to 5 m deep, with hydraulic retention time between 2 and 3 hours. Settled solids (primary sludge) are normally removed from the bottom of tanks by sludge rakes that scrape the sludge to a central well from which it is pumped to sludge processing units. Scum is swept across the tank surface by water jets or mechanical means from which it is also pumped to sludge processing units. Through primary treatment, large particle or inorganic substances are removed from the water. The bod present after primary treatment is 140 mg/L and TSS 130 mg/L. Primary Treatment a) Objective is making the wastewater/sewage compatible for secondary/biological treatment – In case of sewage, primary treatment involves clarification and removal of suspended solids b) Involves generation of (primary sludge) sludge c) Often sludge stabilization is also an objective of primary treatment units d) Usually used primary treatment units – Primary sedimentation/settling tanks or clarifiers – Septic tanks and imhoff tanks – Anaerobic ponds – UASB reactors, UASB ponds and anaerobic baffled reactors e) In some of these units even secondary treatment (biological anaerobic treatment) occurs to a significant extent.

Sedimentation Principles and Types of Settling

Suspended solids present in water having specific gravity greater than that of water tend to settle down by gravity as soon as the turbulence is retarded by offering storage, Inorganic suspended solids having specific gravity of about 2.65 and Organic suspended solids having specific gravity of about 1.04. The particles having specific gravity of about 1.20 or so readily settle down at the bottom of tank. But it is difficult to cause the settlement of lighter particles. This phenomenon of settling down of particles at the bottom of sedimentation tank is known as hydraulic subsidence and every particle has its own hydraulic settling value which will cause its hydraulic subsidence. Basin in which the flow is retarded is called sedimentation basin. Theoretical average time for which the water is detained in the settling tank is called the detention period.

Discrete Settling

TYPE 1 (discrete particle settling): settling of discrete particles in dilute suspensions where concentration is very low, particles have no tendency to flocculate, they settle as individual entities at constant velocity and there is no significant interaction with neighbouring particles and do not flocculate, occurring during pre-settling of sand removal and settling of sand at rapid sand cleaning, Example: removal of grit and sand in wastewater treatment.

If a particle is suspended in water, it initially has 2 forces acting upon it.

1. The forces of gravity,

$$f_g = \rho_p g V_p \tag{12}$$

2. The buoyant force quantified by Archimedes,

$$f_b = \rho_w g V_p \tag{13}$$

Once motion has been initiated; a third force is created due to viscous friction

3. Drag force,

$$f_D = C_D A_p \rho_w \frac{v_s^2}{2} \tag{14}$$

4. Settling velocity of the discrete particle of any shape,

$$v_s = \sqrt{\frac{2g\left(\rho_p - \rho_w\right)V_p}{C_D A_p \rho_w}} \tag{15}$$

For laminar flow,

$$C_D = \frac{24\mu}{v_s d \rho_w} \tag{16}$$

5. (STOKE'S LAW) or Settling velocity of spherical discrete particles under laminar flow conditions, equation expresses the settling velocities of small spherical particles in a fluid medium. Stokes's law finds application in several areas, particularly with regard to the settling of sediment in fresh water and in measurements of the viscosity of fluids. Because its validity is limited to conditions in which the motion of the particle does not produce turbulence in the fluid

$$v_s = \frac{g\left(\rho_p - \rho_w\right)d^2}{18\mu} \tag{17}$$

6. Settling velocity of spherical discrete particles under turbulent flow conditions,

$$v_s = \sqrt{\frac{10g\left(\rho_p - \rho_w\right)d}{3\rho_w}} \tag{18}$$

where,

ρ_P = density of particle,
V_P = volume of particle,
ρ_w = density of water,
g = gravitational acceleration
C_D = drag coefficient,
d = diameter of the particle
A_P = crossection area of particle perpendicular to the direction of the movement,
ρ_w = density of water
v_s = velocity of the particle
μ = kinematic viscosity of water in m^2/s

Example 6

A grit chamber is to be used to remove suspended sand particles from water by allowing them to settle out. The main purpose of this chamber is to prevent the sand from wearing out pipes and pumps within the wastewater treatment plant. To design the tank in a sufficient size, we must first determine the settling velocity of the sand particles. If the sand particles are spheres with diameter dp = 100μm and density 2.65 g/cm3, what would be their settling velocity? (Note that the viscosity of water is 0.11g/cm/sec and the density of water is 1 g/cm3.

Sol. Using Stoke's Law, equation

$$v_s = \frac{\left(2.65gcm^{-3} - 1.0gcm^{-3}\right)\left(980cms^{-2}\right)\left(100\times10^{-4}\,cm^2\right)}{18\times0.1185gcm^{-1}s^{-1}}$$ = 0.76 cm/sec or 27 m/hr

Figure 11. Relationship between drag force (C_D) and Reynolds number (Re)

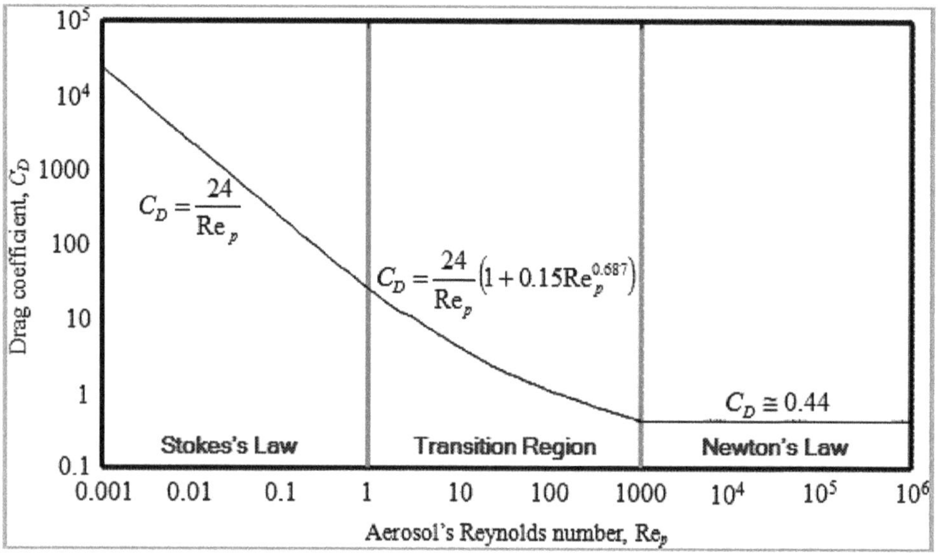

The settling velocity is relatively slow. For very small particles, the drag force is significant in comparison to the gravitational force acting on their small mass.

Example 7

Calculate the gravitational settling velocity for two atmospheric particles having diameters of 0.1µm and 100 µm, respectively, and with density 1 g/cm3. Based on the result, determine whether removal of these particles in a settling chamber of 1 m depth would be a realistic proposition. The viscosity of air is µair = 1.695 X 10-4 g/cm/s. The density of air is extremely small, and thus the buoyancy force may be neglected.

Sol. For the 0.1µm particle,

$$v_s = \frac{\left(1.0gcm^{-3}\right)\left(980cms^{-2}\right)\left(0.1\times10^{-4}\,cm^2\right)}{18\times1.695gcm^{-1}s^{-1}} = 3.2 \text{ x } 10^{-5} \text{ cm/sec}$$

and, for the 100 µm particle,

$$v_s = \frac{\left(1.0gcm^{-3}\right)\left(980cms^{-2}\right)\left(100\times10^{-4}\,cm^2\right)}{18\times1.695\times10^{-4}gcm^{-1}s^{-1}} = 32 \text{ cm/sec}$$

The 0.1 micron particle would take about 900 hours to settle a distance of 1 m. However, because the settling velocity is proportional to the square of the particle size, the 100µm particle settles 106 times faster, and would settle a distance of 1 m in only 3 s. It would clearly be realistic to construct a chamber with a residence time of >3 s. However, a much larger chamber would be required to reach a residence time of 9 hr, and such a chamber would likely not be economical.

Note that the 100 µm particle in this example settles with a velocity that is much greater than that of the 100 µm particle settling through water in example. This is due to the much lower viscosity of air relative to water.

Example 8

A settling chamber is used to remove sand particles from the sewage flow through a wastewater treatment plant. The chamber is 2 m deep and the residence time (retention time) of water in the chamber is 4.4 h. What is the minimum size particle which would be completely removed by settling to the bottom of the chamber during passage through the chamber? The density of sand particles is 2.65 g/cm3, and the viscosity of water is 0.01185g/cm/sec. Assume that any particle that settles to the bottom of the chamber is removed.

Sol. Since only those particles which reach the bottom of the chamber are removed, 100% removal requires that the distance settled during passage through the chamber is equal to the chamber depth. This results in a minimum settling velocity:

$$v_s > \frac{2m}{4.4hr}\times\frac{hr}{3600s}$$

Plugging in our equation for settling velocity,

$$v_s = \frac{\left(980cms^{-1}\right)\left(2.65-1gcm^{-3}\right)}{18\left(0.01185gcm^{-1}s^{-1}\right)}D_p^2 > 1.26\times10^{-4}cm/s$$

Solving for D_p^2

$$D_p^2 > 1.66 \times 10^{-6},$$

or

$$D_p > 1.29 \times 10^{-3} cm = 12.9 \mu m$$

Thus, the minimum size particle removed entirely has a diameter of 13 μm

Flocculent Settling

TYPE 2 (flocculent settling): settling of flocculent particles in dilute suspensions, as particle settle and coalesce with other particles, the sizes of particles are constantly changing and their settling velocity increases with depth and extent of flocculation, having lower in concentration occurring during alum and iron coagulation, Examples: removal of SS in primary sedimentation tanks of WWTP, settling of chemically coagulated waters.

Zone Settling

TYPE 3 (hindered settling) or (zone settling): settling of intermediate concentration of flocculent particles, particles are so close together that interparticle forces are able to hold them in fixed positions relative to each other and the mass of particles settles as a zone at a constant velocity, settle as a mass and form a layer (blanket), concentration is very high greater than 1000 mg/L and the distinct and clear zone is present, occurring at lime softening sedimentation and sludge thickeners in water treatment, Example: biological flocs removal in secondary settling basins of WWTP.

Types of Sedimentation Tanks

Sedimentation tanks can be installed before or after water and wastewater completes the treatment process. Sedimentation Tanks are generally made up of reinforced concrete and may be rectangular or circular in plan. The capacity and other dimension of the tank should be properly designed so as to affect a fairly high percentage of removal of suspended particles. A

Figure 12. Discrete, flocculent and hindered or zone settling

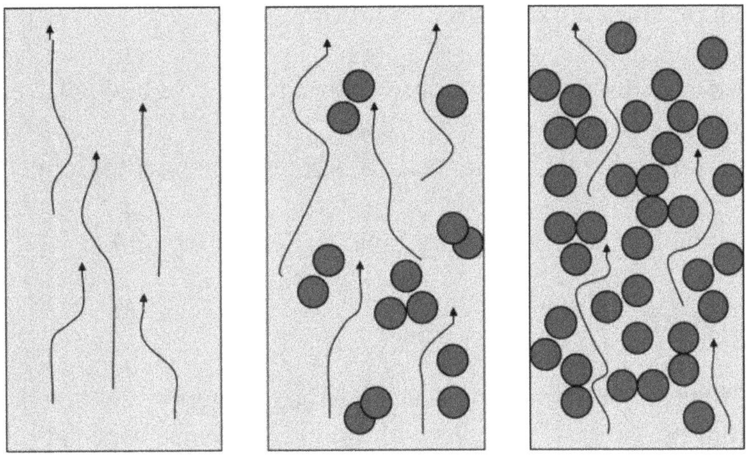

Figure 13. Different levels of settling in the chamber

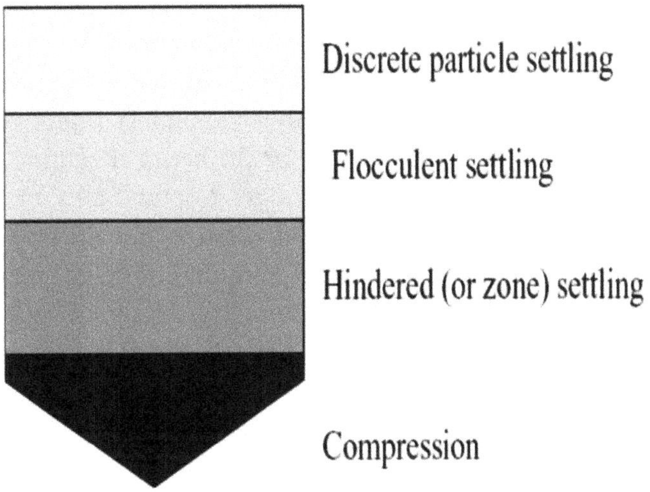

plain sedimentation tank under normal conditions may remove as much as 70% of the suspended impurities present in water. Long narrow rectangular tanks with horizontal flow are generally preferred to the circular tanks with the horizontal radial or spiral flow. Storage reservoirs may also serve as sedimentation basins but they cannot affect proper sedimentation because of factors such as density currents, the turbulences caused by winds etc. and hence they cannot be relied upon. Therefore, special basins are constructed to purify water by the process of sedimentation.

Types of Sedimentation Tanks; Depending upon various factors sedimentation tanks are classified as follows.

1. Based on methods of operation: a) Fill and draw type tank, b) Continuous flow type tank
2. Based on shape: a) Circular tank, b) Rectangular tank c) Hopper bottom tank
3. Based on location: a) Primary tank, b) Secondary tank

Septic Tanks

A Septic Tank is a below ground watertight box (generally contain raw faecal sludge from pit latrines, infected with microorganisms, especially harmful bacterias, which may cause soil and ground water contamination) of brick, precast reinforced concrete or reinforced concrete poured in place, or made available by materials such as composites, HDPE etc., often about 9 x 5 x 6 feet. It separates the liquids and solids, provides digestion of some organics (mainly by bacteria which live without oxygen or anaerobes) and storage. It discharges partially filtered and clarified effluent to the drain field for final treatment. It is a horizontal continuous flow bed tank (with longer detention time), extended sedimentation tank of 12-36 hours detention, closed for anaerobic decomposition of settled sludge. Septic tanks are ordinarily designed for 24 h liquid retention time at average daily flow. Considering the volume required for sludge and scum accumulation, the septic tank may be designed for wastewater retention time of 1 to 2 days. Every septic tank should be provided with the ventilation pipe with the top of the pipe covered with suitable mosquito proof wire mesh. The top of the pipe should extend to at least 2 m above the highest building height present in the vicinity of 20 m from the septic tank.

The ratio of peak flow to average flow may be very high for the small septic tanks, and can disturb the functioning of the tank due to flow surges, leading to washout of the settled solids. The liquid depth of the tank is 1 to 2 m and the length to the width ratio is in the range of 2:1 to 4:1. The sludge accumulated in the tank is cleaned at the frequency of once in 2 to 3 years. Minimum of 300 mm of free board should be provided in the tank. The effluent of the septic tank is offensive and potentially dangerous. Hence, further treatment for septic tank effluent is necessary to protect the receiving environment. Due to inadequate treatment offered to the sewage, septic tanks are recommended for individual houses and for cluster of houses or institutes where contributing population is not exceeding 300 persons. Post treatment can be achieved by aerobic treatment or subsurface disposal. Diffused air aeration with solids recycling (extended aeration), sand filter or synthetic media filter (attached growth process) can be used for treatment of septic tank effluent. Filter bag equipment and hypochlorite addition will also be suitable for treatment. However, frequent replacement of filter bag and hypochlorite addition makes it costly.

Design Features of Septic Tank: The tank should be large enough to provide space for sedimentation of solids, digestion of settled sludge, and storage of sludge and scum accumulated between successive cleaning.

1. **Sewage Flow:** The flow of sewage is considered to be proportional to the number of fixture units discharging simultaneously. One fixture unit is treated as equivalent to the flow of 10 L/min. This is equivalent to the discharge generated from one water closet (WC) when flushed. The number of fixtures discharging simultaneously depends on the population served. For example for the population of 5 persons, number of fixtures will be one and probable peak discharge will be 10 L/min. Similarly for population of 10, 20, and 30 numbers of fixtures will be 2, 3, and 4, and probable peak discharge will be 20 L/min, 30 L/min, and 40 L/min, respectively.
2. **Detention Time:** The detention time of 24 to 48 h is provided for average flow conditions. However, the flow variation is substantial from tank to tank depending upon water usage; and it is not important design criteria.
3. **Sludge Withdrawal:** The sludge is withdrawn at a frequency of 6 months to year in large tank. For small tank it can be 2 to 3 years.
4. **Capacity of the Tank:** The total capacity of the septic tank is worked out using following considerations.
    a. **Sedimentation:** An area of 0.92 m^2 is required for every 10 L/min peak flow rate to support adequate sedimentation of suspended solids. This will favour sedimentation of solids with 0.05 mm size and sp. gravity of 1.2. A minimum of 0.25 to 0.30 m depth is necessary for sedimentation.
    b. **Sludge Digestion:** The SS per capita may be considered as 70 g/day. It is assumed that that 60% of the solids will be removed in the tank, out of which 70% solids will be volatile, with 5% solid content i.e., 95% water content.
    c. The volume of fresh sludge = 0. 84 L/Capita/day, Considering that 2/3 of the volatile matter is destroyed of which ¼ is mineralized

during digestion and solids content of 13% in digested sludge, the volume of total digested sludge, i.e., mineralized sludge with 13% solids plus undigested sludge with 5% solids, will be 0.234 L/Capita-day. The digestion zone contains both the fresh sludge (which is simultaneously getting destroyed by 2/3 of its volume) and digested sludge; hence volume of both of these will work out to be (0.848*1/3+0.234) = 0.516 L/Capita-day. At 25° C the typical time required for digestion will be 63 days. Hence, capacity of digestion zone works out to be 63 * 0.0005 = 0.032 m^3/capita.

d. **Volume Required for Sludge and Scum Storage:** For interval of 1 year of sludge cleaning, sludge storage capacity of 0.0002*365 = 0.073 m^3/cap is required. The 25 to 50 mm of seed volume should be considered, and care should be taken while withdrawing the sludge to leave this volume of sludge to act as seed. No separate depth is provided for this. Total Capacity: Hence the total capacity of the septic tank will be equal to sum of the above three requirements, plus a minimum free board of 0.3 m should be provided. Therefore for 20 persons the total capacity of the septic tank will be

5. **Sedimentation:** Considering peak flow of 30 L/min for 20 persons, the area required = 0.92*30/10 = 2.76 m^2. Keeping depth of min. 0.3 m for sedimentation, the volume = 2.76*0.3 = 0.828 m^3
6. **Digestion:** 0.032 * 20 = 0.64 m^3
7. Sludge storage: 0.073 * 20 = 1.46 m^3 for one year. For 2 year cleaning frequency sludge storage volume required = 1.46 * 2 = 2.92 m^3
8. Free board = 2.76 * 0.3 = 0.828 m3. Hence, total volume of septic tank for, 20 person = 0.828 + 0.64 + 2.92 + 0.828 = 5.216 m^3.

Height of the septic tank = 0.3 + 0.231 + 1.05 + 0.3 m = 1.881 m, and provide length to the width ratio of 3; hence L = 2.88 m and W = 0.96 m.

Two-Story (Imhoff) Tank

The Imhoff tank is a primary treatment technology for raw wastewater, designed for solid-liquid separation and digestion of the settled sludge. It consists of a V-shaped settling compartment above a tapering sludge digestion chamber with gas vents. In the digestion chamber, the settled solids are anaerobically digested generating biogas. The gas is deflected by baffles to the gas vent channels to prevent it from disturbing the settling process. Imhoff tanks are used by small communities and due to the underground construction, land use is very limited. Investment costs are low and operation and maintenance simple. But the treatment efficiency is low and a secondary treatment of the effluent is required. The Imhoff tank (also known as Emscherbrunnen or Emscher Tank), which works similar to a communal septic tank, is a robust and effective settler that causes a suspended solids reduction of 50 to 70%, COD reduction of 25 to 50%, and leads to potentially good sludge stabilisation – depending on the design and conditions. It is a compact and efficient system for pre-treatment of municipal wastewater. The settling compartment has a circular or rectangular shape with V-shaped walls and a slot at the bottom, allowing solids to settle into the digestion compartment, while preventing foul gas from rising up and disturbing the settling process. Gas produced in the digestion chamber rises into the gas vents at the edge of the reactor. It transports sludge particles to the water surface, creating a scum layer. The sludge accumulates in the sludge digestion chamber, and is compacted and partially stabilised through anaerobic digestion. The liquid fraction remains only some hours in the tank, while the solids remain for several years in the digestion chamber. There is more biogas production than in septic tanks, but unfortunately the organic load in communal wastewater is usually not high enough for the economical collection and usage (or flaring) of biogas (Hoffmann et al., 2011). The pre-treated wastewater from the Imhoff tank requires a secondary treatment (e.g. leach field, soak pits, horizontal flow, vertical flow or free-surface constructed wetlands). Also the faecal sludge needs to be correctly disposed and further treated (e.g. small or large scale composting, settling - thickening ponds or drying beds). If the sludge is composted either directly or after drying, it can be used as fertiliser to improve the soil quality.

Design Considerations: The Imhoff tank is designed with three compartments (see picture below):

1. Upper compartment for sedimentation
2. Lower section for sludge digestion
3. Gas vent and scum section

Scum and gas vent chambers are located at the sides of the tank. It can be an open or covered tank. A covered tank avoids infiltration of precipitation and pollutants. However, it is always preferred to cover the tank to prevent contact with people and animals. An outlet for de-sludging can be added.

The Imhoff tank is usually built underground with reinforced concrete. It can, however, also be built above ground, which makes sludge removal easier due to gravity, although still requiring pumping up of the influent. Small prefabricated Imhoff tanks are also available on the market. Hydraulic retention time is usually not more than 2 to 4 h to preserve an aerobic effluent for further treatment or discharge. T-shaped pipes or baffles are used at the inlet and the outlet to reduce velocity and prevent scum from leaving the system. The total water depth in the tank from the bottom to the water surface may reach 7 to 9.5 m. The bottom of the settling compartment is typically sloped 1.25 to 1.75 vertical to 1 horizontal and the slot opening can be 150 to 300 mm wide. The walls of the sludge digestion compartment should have an inclination of 45° or more. This allows the sludge to slide down to the centre where it can be removed. Dimensioning of the anaerobic digestion compartment depends mainly on sludge production per population equivalent, on the targeted degree of sludge stabilisation (linked to the de-sludging frequency) and the temperature. The digestion chamber is usually designed for 4 to 12 months sludge storage capacity to allow for sufficient anaerobic digestion. In colder climates longer sludge retention time and, therefore, a greater volume is needed. For de-sludging, a pipe and pump have to be installed or access provided for vacuum trucks and mobile pumps. A bar screen or grit chamber is recommended before the Imhoff tank to prevent coarse material from disturbing the system.

Construction costs are slightly higher than the costs for a septic tank. Also the costs for de-sludging (motorised or manual) must be considered. Moreover, it is a pre-treatment facility and in many cases connected to further treatment installations (e.g. leach field, soak pits, horizontal flow, vertical flow or free-surface constructed wetlands). Imhoff tanks are recommended for domestic or mixed wastewater flows between 50 and 20,000 population equivalents. They are able to treat high organic loads and are resistant against organic shock loads, Low Space requirements. Normally, Imhoff tanks are used by small communities with raw wastewater flows on the order of 950 m^3/day (population about 8000 people or 1300 households), Imhoff tanks can be used in warm and cold climates (wastewater temperatures below 15°C or above 2000 m altitude; minimum winter temperature is 8°C, average for the year is 20°C). As the tank is very high, it can be built underground if the groundwater table is low and the location is not flood prone. If constructed underground, land use is very limited and Imhoff tanks can be constructed in both, rural or urban areas. Investment costs are low and operation and maintenance simple.

Plain Settling Tanks (Clarifiers)

Plain Sedimentation

It is process of settling down of solids and impurities in the raw water to the bottom of sedimentation basin by natural gravity force alone, no chemical is added. This method is very cheaper and mostly used in all filtration and purification system of water. Sedimentation by using clarifier and contact: In this method chemicals are mixed in water and that water is rotated by help of pumps for period of two hours per day, and suspended solids are settled down in the bottom of reservoir or tank etc.

Chemically assisted sedimentation or clarification: this is process in which chemicals are added to water and through mixing the suspended solids and other impurities are stick together and form flocs, which settles to the bottom of basin. Sedimentation tank may function either intermittent or continuously. The intermittent type tank is those which store water for a certain period and keep it on completely rest. In continuous type tank, the flow velocity is only reduced and water is not brought to complete rest. Sedimentation tank or basin may be either circular or rectangular in plan. Long narrow rectangular tanks with horizontal flow are generally preferred to the circular tank of radial or spiral flow.

Rectangular sedimentation basin: Rectangular basins are the simplest design, allowing water to flow horizontally through a long tank this type of basin is usually found in large-scale water treatment plants. Rectangular basins have a variety of advantages such as predictability, cost-effectiveness, and low maintenance. They are the least likely to short-circuit, especially if the length is at least twice the width. The inlet and outlet arrangements of rectangular basis are shown below. A disadvantage of rectangular basins is the large extent of land area required. Circular basin: Square and circular sedimentation basins with horizontal flow are often known as clarifiers. This type of basin faces short-circuiting problems. Hopper bottom tank: The shape of the base of this tank is hopper type. This is vertical flow tanks, because water flows upward & downward in this tank. The water enters in this tank from top inlet channel. Because of deflector box water flows from upper to lower, impurities settled at the bottom of the tank, pure water taken by draw off channel. Sludge outlet pipe is used to take out the sludge from tank. Detention period for plain sedimentation is 4-8 hours and for coagulated sedimentation 3-4 hours, the velocity of flow not greater than 30 cm/min (horizontal flow). Tank dimensions: L:B = 3 to 5:1. Generally L= 30 m (common) maximum 100 m, Breadth= 6 m to 10 m. Circular: Diameter not greater than 60 m, generally 20 to 40 m, Depth 2.5 to 5.0 m (3 m). Surface Overflow Rate: For plain sedimentation 12000 to 18000 L/d/m^2 tank area; for thoroughly flocculated water 24000 to 30000 L/d/m^2 tank area. Slopes: Rectangular 1% towards inlet and circular 8%.

Settling Operations

Particles falling through the settling basin have two components of velocity:

1. Vertical component: $v_v = \dfrac{g\left(\rho_p - \rho_w\right)d^2}{18\mu}$
2. Horizontal component: $v_h = \dfrac{Q}{A}$

The path of the particle is given by the vector sum of horizontal velocity v_h and vertical settling velocity v_t

Figure 14. Schematic design of (a) rectangular and (b) circular basin

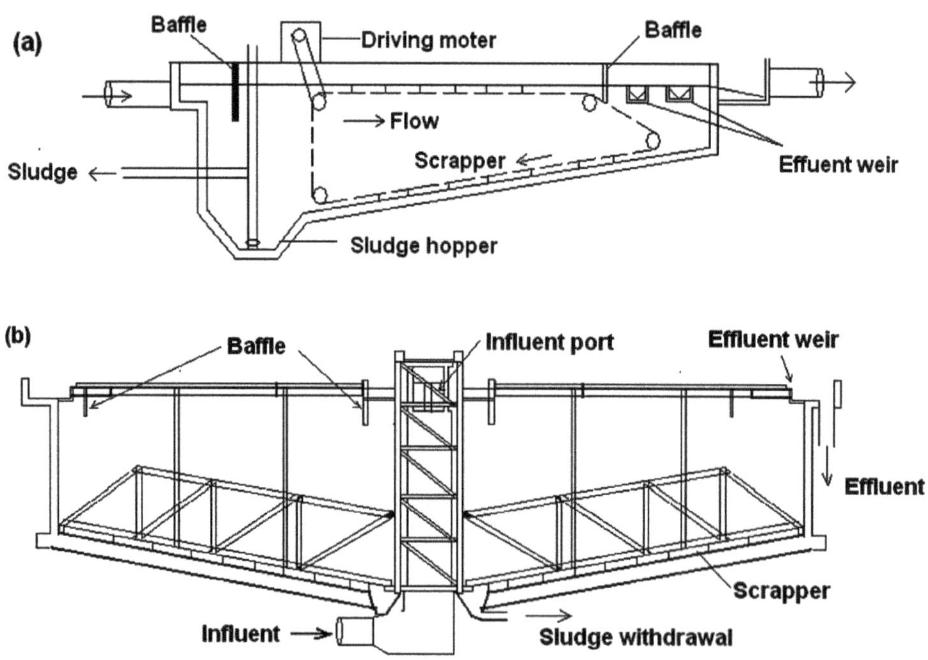

Assume that a settling column is suspended in the flow of the settling zone and that the column travels with the flow across the settling zone. Consider the particle in the batch analysis for type-1 settling which was initially at the surface and settled through the depth of the column Z_0, in the time t_0. If t_0 also corresponds to the time required for the column to be carried horizontally across the settling zone, then the particle will fall into the sludge zone and be removed from the suspension at the point at which the column reaches the end of the settling zone.

All particles with $v_t > v_0$ will be removed from suspension at some point along the settling zone.

Now consider the particle with settling velocity $< v_0$. If the initial depth of this particle was such that Z_p/v_t=t0, this particle will also be removed. Therefore, the removal of suspended particles passing through the settling zone will be in proportion to the ratio of the individual settling velocities to the settling velocity v_0. The time t0 corresponds to the retention time in the settling zone,

$$t = V/Q = LZ_0W/Q,$$

$$\text{Also } t_0 = Z_0/V_0 \tag{19}$$

$$\text{Therefore, } Z_0/V_0 = LZ_0W/Q$$

$$\text{And } V_0 = Q/LW$$

$$\text{Or } V_0 = Q/A_s \tag{20}$$

Thus, the depth of the basin is not a factor in determining the size particle that can be removed completely in the settling zone. The determining factor is the quantity Q/A_s, which has the units of velocity and is referred to as the overflow rate q_0. This overflow rate is the design factor for settling basins and corresponds to the terminal setting velocity of the particle that is 100% removed.

PROCESS CONTROL CALCULATIONS

Percent Removal

Both organic and inorganic solids are present in wastewater, and both can be either suspended or dissolved. Settleable solids are the portion of suspended solids that readily settle in a primary sedimentation tank when the wastewater velocity is reduced to a fraction of a meter or foot per second. Colloidal solids, which are finely divided solids, are too fine to settle within the usual detention times of a primary sedimentation tank. Colloidal solids readily pass through the primary treatment process and are treated in the secondary treatment process. Primary sedimentation tanks reduce the wastewater velocity to less than 0.3 m/s (1.0 ft/sec) and allow these settleable solids to separate from the waste stream. This process also removes a percentage of suspended solids as well as Biochemical Oxygen Demand (BOD) that are associated with these solids. Typical removal efficiencies that can be achieved in primary treatment are Settleable Solids 90 – 95%, Suspended Solids 50 – 65%, BOD 20 – 35%.

Detention Time

The time of wastewater stays in the settling tank is called the detention time or retention time. Approximately 1.5 – 2.5 (2.0) hours of detention time are needed in the primary settling tank. The exact time depends on many factors such as the influent flow rate and the removal requirements needed by downstream processes. If the detention time is too long, solids may become septic and float to the surface. High suspended solids levels in the primary effluent and subsequent odours may result. A secondary clarifier requires a longer detention time than a primary settling tank because the light and fluffy activated sludge particles do not settle as easily as the heavier solids removed in a primary tank. How efficiently the settling tank removes settleable solids depends on how slow the liquid moves (influent velocity) and on the detention time.

$$Detention\,Time\left(\theta\right) = \frac{Volume\,of\,the\,Primary\,Se\dim antation\,Tank}{Influent\,Flow\,Rate} \tag{21}$$

Surface Loading Rate (Surface Settling Rate, Surface Overflow Rate)

The surface overflow rate is a measure of how rapidly wastewater moves through the settling tank. When we talk about surface overflow rate, we are referring to the number of gallons going through the settling tank each day for each square foot of surface area in the tank, or the number of liters for each square meter per day. In other words, we are looking at the hydraulic wastewater load for each square meter, or square foot, of surface area in the settling tank each day (L/m^2/day) 25,000 – 50,000 L/m^2/day for Primary settling only and for Primary settling with Secondary treatment 35,000 – 80,000 L/m^2/day. The surface overflow rate is defined as the loading across the surface of the primary tank defined as follows:

$$Surface\,Over\,Flow\,Rate = \frac{Influent\,Flow\,Rate}{Surface\,Area\,of\,the\,Tank} \quad (22)$$

Weir Overflow Rate (Weir Loading Rate)

Wastewater leaves the settling tank, collected into a collection by flowing over weirs and into effluent troughs or launders, the purpose of a weir is to allow a thin film of the clearest water to overflow the tank so the material or settled suspended solids should not be washed out with the clarified effluent, A high velocity near the weir can pull settling solids into the effluent. Weir length is designed to avoid washout out of sludge Rectangular Primary Sedimentation Tanks. The length of the weir in the settling tank compared to the flow is important in preventing high velocities. A baffle at the outlet end of a rectangular tank or around the edge of a circular tank helps prevent short-circuiting and floating solids from leaving the tank. Baffles are also used near the outlet weirs to help deal with density currents. Two of the more common types used are the Crosby and Stamford peripheral baffles. Most tank weirs can be adjusted and made level so that effluent flow is uniformly distributed. Assuming that flow over the weir is uniformly distributed, one way to determine the sufficient weir length is to calculate the daily flow over each meter, or each foot, of weir. This measurement is called the weir overflow rate. The weir overflow rate equals the number of liters per meter of weir per day, or the number of gallons of wastewater that flows over one foot of weir per day (Primary treatment).

$$Weir\,Over\,Flow\,Rate = \frac{Influent\,Flow\,Rate\left(L\,/\,day\right)}{Weir\,Length\left(m\right)} \tag{23}$$

Sludge Pumping

Another important step in the settling process is sludge removal. Since the main purpose of a primary settling tank is to allow solids to settle out of the wastewater, we cannot just leave them in the tank. The de-sludging of settling tanks and clarifiers has always been a very costly and timely process. Often plants hire expensive vacuum trucks which take a lot of time to suck out all the sludge. This is also a dangerous option, as it exposes personnel to toxic material located in the settling tanks. Other options are to utilize centrifugal pumps, which pump very low percent of solids and often get clogged by the debris. Sludge's with less than 10 percent solids can be pumped through force mains. Sludge's with solids contents less than 2 percent have hydraulic characteristics similar to water. For solids contents greater than 2 percent, however, friction losses are from 1 1 / 2 to 4 times the friction losses for water. Both head losses and friction increase with decreasing temperature. Velocities must be kept above 2 feet per second. Grease content can cause serious clogging, and grit will adversely affect flow characteristics as well. Adequate clean-outs and long sweep turns will be used when designing facilities of these types.

In a rectangular tank cross section including the solids removal equipment, the main components are the flights, drag chains, head shaft and idler shafts. The wooden or fiberglass beams, commonly called flights, are attached to drag chains, which are connected to form a closed loop. The head shaft is rotated by the drive chain. This rotation causes the drag chains and flights to move through the settling tank. Solids that settle to the bottom of the settling tank are scraped to a hopper or trough. Most small rectangular tanks have two hoppers. Solids collected in these hoppers must be removed. Larger settling tanks usually have a trough running the entire width of the tank. In this type of system, scrapers are used to move the solids to one end of the trough for removal. This is called a cross-collector. In the circular settling tank, scrapers called plows, move solids into a hopper at the centre of the tank. These plows are driven by a motor mounted above the feed well structure. In both circular and rectangular tanks, solids are moved very slowly so that they are not mixed and suspended in the wastewater again. After settled sludge has been moved to the sludge hopper, it still has to be removed completely from the tank. The method used to remove this sludge will affect the sludge stabilization process. For example, if plant uses anaerobic digesters, the smaller the volume of sludge that pump into the digester. Because most plants' digesters are built to handle only the minimum volume necessary for continuous treatment, it is important to pump sludge wisely. All sludge must be removed from the primary tanks, so it should be concentrated into the least possible volume. This means pumping the sludge with as little water as possible. The solids collected in the primary tank hopper are pumped to the sludge stabilization process or solids handling process. What happens to the primary sludge will depend on the plant design. Solids handling systems vary from plant to plant and include the use of aerobic digesters, anaerobic digesters, centrifuges, belt presses, and other solids handling processes.

The amount of sludge pumped from the primary tanks is an important factor, and the type of equipment used to remove the sludge varies. Typically, treatment plants use piston pumps, diaphragm pumps, or progressing cavity pumps to remove sludge from primary tanks. Some plants use centrifugal-type pumps. However, the capacity of centrifugal pumps can be affected by the solids concentration and sludge characteristics. Many primary sludge-pumping systems have variable pump speed capability, such as manually adjusted belts, variable-frequency drives, or adjustable-gear units. Adjustable pump outputs reduce the chance of coming in the sludge hopper and subsequent pumping of water only. Also, adjusting the pump rates can benefit the solids-handling facilities. Primary sludge-pumping systems typically have start and stop timers. Some plants use timers to start the pumping system and density meters to stop the pumps. Many plants today use programmable computers on their sludge-withdrawal systems, while others use manual timing operations.

Percent Total Solids (%TS)

The percentage of the original mass of a sample that remains after water is evaporated from the sample in a laboratory drying oven at 103°C. So, if 20 g of sample were placed in an evaporating dish of known weight and, after drying, the sample weight 1.0 g more than the dish, the sample would have had 5.0% TS. The quantity of solids is determined because they are important in the control of biological and physical wastewater treatment processes and to assess the compliance with regulatory wastewater effluent limitations. The total solids (TS) are defined as the material residue left in the vessel after evaporation of the sample and drying in the oven at a well-defined temperature. One of the most commonly used methods is the total suspended solids (TSS). These are the solids that are retained by a filter. The volatile suspended solids (VSS) are defined as the weight loss after ignition. To determine the TSS a well-mixed sample is filtered through a weighed glass-fibre filter and the residue retained on the filter is dried to a constant weight at 103 to 105 °C. The increase of weight of the filter represents the TSS.

$$TSS\left(\frac{mg}{L}\right) = \frac{\left(Final\left(Weight\ of\ Filter + Dry\,\mathrm{Re}\,sidue\right) - Initial\left(weight\ of\ filter\right)\right)\left(mg\right)}{Added\ Sample\ Volume\left(ml\right)} \times 100 \tag{24}$$

For volatile solids, a well-mixed sample is first evaporated in a weighed dish and dried in an oven at 103 to 105 °C. After heating the increase in weight of the dish is noted and then the residue is ignited at 550 °C. The remaining solids after the ignition represent the mass of fixed total solids and the weight loss due to ignition represents the mass of the volatile suspended solids. The determination of the fixed and the volatile suspended solids are useful as a control for the operations of a wastewater treatment plant because it offers an approximation of the amount of organic matter present in the solid fraction of the wastewater and activated sludge.

$$VSS\left(\frac{mg}{L}\right) = \frac{\begin{pmatrix} Final\left(Weight\ of\ Dish + Residue\ After\ Ingnitian\right) \\ -Initial\left(Weight\ of\ Dish\right) \end{pmatrix}\left(mg\right)}{Added\ Sample\ Volume\left(ml\right)} \times 100 \tag{25}$$

Bod and SS Removal

Suspended solids, in addition to contributing to BOD, may settle on the stream bed and inhibit certain forms of aquatic life. The BOD if discharged into a stream with low flow can cause damage to aquatic life by reducing the dissolved oxygen content. In addition the secondary effluent contains significant amounts of plant nutrients and dissolved solids. If the waste water is of industrial origin, it may also contain traces of organic chemicals, heavy metals and other contaminants. Of Suspended Solids: This treatment implies the removal of those materials that have been carried over from a secondary treatment settler. Many methods were proposed of which two methods were commonly used. The two methods are micro staining and chemical coagulation followed by settling and mixed media filtration. a) Micro straining: It is a special type of filtration procedure which makes use of filters oven from stainless steel wires with opening only 60-70 μm across to remove very small particles. High flow rates and low back pressures are normally achieved. b) Coagulation and flocculation: The object of coagulation is to alter these particles in such a way as to allow them to adhere to each other. Most colloids of interest in water treatment remain suspended in solution because they have a net negative surface charge that causes the particles to repel each other. The intended action of the coagulant is to neutralise that charge, allowing the particles to come together to form larger particles that can be more easily removed from the raw water. The usual coagulant is alum $[Al_2(SO_4)_2 \times 18H_2O]$.

Though FeC13, FeSO4and other coagulants, such as polyelectrolytes, can be used. Alum when added to water, the aluminium in this salt hydrolyses by reactions that consume alkalinity in the water such as:

$$[Al(H_2O)_6]^{3+} + 3HCO_3^- \rightarrow Al(OH)_3(s) + 3CO_2 + 6H_2O$$

The gelatinous hydroxide thus formed carries suspended material with it as it settles. In addition, however, it is likely that positively charged hydroxyl-bridged dimers, and higher polymers are formed which interact specifically with colloidal particles, bringing about coagulation. Metal ions in coagulants also react with virus proteins and destroy up to 99% of the virus in water. Anhydrous ion (III) sulphate can also act as effective coagulant similar to aluminium sulphate, an advantage with iron (III) sulphates it that it works over a wide range of pH. c) Filtration: If properly formed, the addition of chemicals for promoting coagulation and flocculation can remove both suspended and colloidal solids. After the flocs are formed, the solution is led to a settling tank where the flocs are allowed to settle. While most of the flocculated material is removed in the settling tank, some flocs do not settle. These flocs are removed by the filtration process, which is usually carried out using beds of porous media such as sand or coal. The current trend is to use a mixed-media filter which consists of fine garnet in the bottom layer, silica sand in the middle layer and coarse coal in the top layer which reduces clogging.

The organic matter present in the wastewater may belong to two groups: Carbonaceous matter, Nitrogenous matter, the ultimate carbonaceous BOD of a waste is the amount of oxygen necessary for microorganisms in the sample to decompose the biodegradable carbonaceous material. This is the first stage of oxidation and the corresponding BOD is called as first stage BOD. In the second stage the nitrogenous matter is oxidized by autotrophic bacteria, and the corresponding BOD or nitrification demand. In fact, polluted water will continue to absorb oxygen for many months, and it is not practically feasible to determine this ultimate oxygen demand. Hence the 5 days period is generally chosen for the standard BOD test, during which oxidation is about 60 to 70% complete, while within 20 days period oxidation is about 95% to 99% complete. A constant temperature of 20°C is maintained during incubation. The BOD value of 5 Day incubation period is commonly written as BOD_5 or simply as BOD. Another reason for selecting 5 days as standard duration is to avoid interference of nitrification bacteria. Nitrification starts after 6th or 7th day. Sanitary engineers are generally interested in carbonaceous BOD only, so by selecting 5 days we generally get only the carbonaceous BOD. Interference of Nitrification can be eliminated by pre-treatment of sample or by using inhibitory agents like methylene blue. Now a day BOD test is also done at 27°C and duration of 3 days (BOD_3), results can be obtained faster and it is more nearer to the actual field conditions.

$$BOD = \left(Initial\,D.O. - Final\,D.O.\right) \times Dilution\,factor \tag{26}$$

$$BOD = \left(Initial\,D.O. - Final\,D.O.\right) \times \frac{Volume\,of\,the\,diluted\,sample}{Volume\,of\,the\,undiluted\,sewage\,sample} \tag{27}$$

Here the dilution factor means, for example: 1% dilution sample means, 1 ml of sewage sample is diluted to make 100 ml of test sample, hence the dilution factor is 100, as 1 ml has been diluted by 100 times to make 100 ml. Hence the multiplied dilution factor would be 100. In actual the BOD bottles are of 300 ml volume, which is commonly used for the experiment. The given volume of sample, say 5 ml placed in the bottles and mixed with pure aerated dilution water to make 300 ml diluted sample. These bottles then incubated at 20°C for 5 days. Light must be excluded from the incubator to prevent the algal growth that may produce oxygen in the bottles. The D.O content before and after incubation then determined. The BOD of the sewage is then calculated as:

$$BOD = \left(Initial\ D.O. - Final\ D.O.\right) \times \left(\frac{300}{5}\right)$$

Limitations of BOD test are; it measures only the biodegradable organic matter. Time duration of the test is very long i.e. 5 days, so if quick results are needed it is not useful. Pre-treatment is needed if the sample contains toxic waste. Nitrifying bacteria can cause interferences and could give higher results. To avoid them proper care must be taken. It is essential, to have high concentration of active bacteria present in the sample.

Examples 9

The BOD_6 of a wastewater is determined to be 400 mg/L at 20° C. The k value at 20° C is known to be 0.23 per day. What would be BOD_8 value if tests were run at 15° C?

Sol. The BOD can determined by using the formula given below

$$BOD_{t,T} = BOD_u\left(1 - \exp^{\left(-K_T \times t\right)}\right) \tag{28}$$

$$K_T = K_{20}\left(\theta\right)^{\left(T-20\right)} \tag{29}$$

$$K_T = K_{20}\left(1.047\right)^{\left(T-20\right)} \tag{30}$$

Given: $BOD_{6,\ 20} = 400$ mg/L, $K_{20} = 0.23\ d^{-1}$

$$BOD_6 = BOD_u\left(1 - \exp^{\left(-20 \times 6\right)}\right)$$

$$BOD_u = 534.458 mg / L$$

$$K_{15} = K_{20}\left(1.047\right)^{(15-20)}$$

$$K_{15} = 0.183 \text{ d}^{-1}$$

Now,

$$BOD_{8,15} = 534.458\left(1 - \exp^{(-K_{15}\times 8)}\right)$$

$$BOD_{8,15} = 410.643 mg / L$$

Examples 10

6 ml of wastewater is diluted to 300 ml distilled water in standard BOD bottle. Initial DO in the bottle is determined to be 8.5 mg/l. DO after 5 days at 20 C is found to be 5 mg/l. Determine BOD_5 of wastewater and compute the ultimate BOD.

Sol. Given,

$$BOD = \left(Initial\ D.O. - Final\ D.O.\right) \times \frac{Volume\ of\ the\ diluted\ sample}{Volume\ of\ the\ undiluted\ sewage\ sample}$$

$$BOD = \left(8.5 - 5\right) \times \frac{300}{6}$$

$$BOD = 256 mg / L$$

PROBLEM ANALYSIS

Causal Factors for Poor Suspended Solids Removal (Primary Clarifier)

The removal of SS results in a significant decrease of organic load, usually expressed in terms of biochemical oxygen demand (BOD) or chemical oxygen demand (COD). When wastewater is placed in a cone (such as an Imhoff cone) and allowed to sit, settleable solids settle to the bottom, and lighter floatable solids rise to the top, This is essentially the same thing that happens in a primary settling tank (sedimentation) cause poor suspended solids removal in the tank. Colloidal, or finely divided, solids that will not settle and dissolved solids will remain in the liquid and be carried on for further processing. If visual observations, confirmed by lab analysis indicate poor clarifier operation, then the problem source must be identified so that corrective and preventive action can be taken.

Some problems related to suspended solids are; Floating Sludge: sludge decomposing in tank and floating to surface, excessive sedimentation in inlet channels due to low velocity for cross-sectional area of channel at existing flow rates can be managed by reduce cross-sectional area of the channel - Agitate with air, water, or other means to prevent deposits. If the speed of the wastewater is greater in some areas of the tank than others, a condition called "short-circuiting" can occur. In places where the wastewater is moving faster, particles that are suspended in the wastewater may not have a chance to settle out. They will be held in suspension and will pass through to the discharge end of the tank. It is desirable to maintain even flow distribution to prevent short-circuiting in the settling tank, the overflow weirs must be perfectly level to ensure good flow distribution and help prevent short-circuiting, and due to improper inlet distribution it retains a higher velocity than the rest of the contents. This newly formed current will simply deflect off of the sludge blanket and use its momentum to carry itself to the clarifier outlet structure, often carrying sludge from the blanket with it. Baffles may be installed near the outlet weirs to help prevent this solids loss. Any solids that float to the surface are removed by scum collection devices and further processed.

Most organic settleable solids having the specific gravity slightly equal to the water. So they settle very slowly. Settling tanks are designed with this fact in mind. The velocity of the liquid in the settling tank is slowed down to a fraction, approximately 0.001 m/s (0.003 ft/sec), of its influent velocity as compared to about 0.3 m/s (1.0 ft/sec) in the grit chamber, and at least 0.6 m/s (2.0 ft/sec) in the sewer. The wastewater must stay in the settling tank long enough for solid particles to settle. If the tank is too small for the volume of flow entering it, too many particles will exit with the tank effluent. Another factor is the sludge with poor settling properties, which results from massive growth of filamentous bacteria, is described as bulking sludge or as foaming sludge respectively, if foam is formed on the water surface. The reasons for formation of bulking- and foaming sludge are not completely clarified and there are no methods to fight these sludge's, which are generally effective, because of the variety of filamentous microorganisms. In many cases this requires a rebuilding and reorganisation of the process layout. From the implementation of the method to the solution of the problem a lot of time may pass and success cannot be guaranteed.

Causal Factors for Floating Sludge

Having floating sludge is sludge forms a very dense thick layer on the reaction tank and the clarifier. Secondary clarifiers can be running smoothly one day and then suddenly solids begin to float and carry over the weir into the effluent known as floating sludge. The floating sludge in the clarifier caused; De-nitrification – small nitrogen gas bubbles float the sludge in the clarifier creating floating sludge chunks with small bubbles entrapped and the problem can be often controlled by increasing the recycle pump to reduce sludge blanket depth/sludge retention time in the clarifier. The problem is often related to an increase in influent ammonia/nitrogen that is converted by beneficial nitrifiers into NO_2 or NO_3 via the autotrophic nitrification process. Without an anaerobic/anoxic step that removes NO_2 or NO_3 in the treatment system, this process occurs in the clarifier. Long run solutions include evaluating influent TKN/ammonia, anoxic de-nitrification zone residence time, availability of "food" or easily available BOD in the anoxic zone for de-nitrification, and an overall system survey on sludge age, residence times, and influent makeup., fats, Oils & Grease – simply put, FOG floats on water. When entrapped in floc, excessive grease or oil can cause floating biomass. This appears as a scum blanket that can cover the entire clarifier.

The FOG created floating sludge involves a messy control process. First the scum/floating sludge needs to be removed or allowed to carry over the weir to polishing/tertiary treatment.

Upstream, operators need to evaluate where the FOG increase originated. This can be a one-time slug or increased loadings of grease over time. It is best to prevent oils and grease from entering the biological treatment system. In cases where we have high levels of FOG in a system we encourage operators to increase wasting rates (remove entrapped FOG this way) and add cultures associated with FOG degradation/bio-surfactant production. By using wasting and seeding steps together, the potential for significant biomass reduction is prevented while removing entrapped FOG that causes high effluent solids, viscous bulking or billowing sludge – viscous bulking can sometimes create floating sludge (more often it is just billowing over the weir versus floating). This is often caused by nutrient deficiencies (normally low phosphate) in industrial waters. The solution here is to evaluate changes in influent makeup and changes in the environmental conditions in the biological treatment unit. While researching the exact cause of the bulking, operators need to begin wasting the bulking, viscous sludge. If nutrient residuals are low (<1.0 mg/L ammonia nitrogen or <0.5 mg/L ortho-phosphate) then begin adding nutrients to achieve residuals above the targets above. If heavy wasting is involved, we recommend adding cultures to promote the shift to a desirable biomass.

Causal Factors for Septic Wastewater or Sludge

Primary tank should never have septic sludge floating on the surface of the primary settling tank. Septic sludge floating on the surface of the water in the tank will be large and inelegant in nature. Dealing with primary sludge is an important aspect of primary treatment operation of the wastewater treatment and in the handling or disposal of sludge. How long sludge stays at the bottom hopper of the primary tank is very important. If the sludge stays in the hopper of the tank for too long time, it will become septic. Gases produced from the septic sludge will cause the sludge to float to the top. Having odour problem are also indicates that the sludge are not pumping from the primary tank or not pumping often enough. When sludge turns septic, hydrogen sulphide (H_2S) is produced. If the rotten smell observed around the primary tank, then make sure that the sludge are not letting stay at the bottom of the tank too long. Besides from causing operational problems, septic sludge can be dangerous. When sludge turns septic, hydrogen sulphide (H_2S) and methane

(CH_4) gases are produced. These gases are dangerous and also flammable, especially in confined spaces. The right mixture of methane to air can cause an explosion. Hydrogen sulphide produced by septic sludge can also change into sulphuric acid (H_2SO_4) can destroy and corrode the concrete in tanks. Therefore, it is important that the sludge does not stay at the bottom of the primary tank for too long. Too much sludge can be pumped from a primary tank, because the sludge will be too thin. However, if the plant uses primary sludge de-gritting or hydro-cyclones, then it is must important to pump primary sludge continuously because hydro-cyclones require very thin sludge (less than 1% solids).

Causal Factors for Too Low Primary Sludge Solids Concentration

Primary sludge accounts for 50-60% of the suspended solids applied. Carbon dioxide, hydrogen sulphide, and odourous organic compounds are liberated from the settlings in the primary basin as a result of anaerobic decomposition, and the solids concentration is rarely greater than 4%. Basic floc formation, required for activated sludge operation due to the use of gravity clarifiers, is due to a growth form of many species of natural bacteria. Floc-forming species share the characteristic of the formation of an extracellular polysaccharide ("slime") layer, also termed a glycocalyx. This material, which consists of polysaccharide, protein and sometimes cellulose fibrils, "cements" the bacteria together to form a floc. Floc formation occurs at lower growth rates and at lower nutrient levels, essentially starvation or stationary growth conditions. Floc-forming species may grow in a dispersed and non-settleable form if the growth rate is too fast. This latter condition, termed dispersed growth, occurs rarely in domestic waste activated sludge operation but occurs often in industrial waste treatment, generally due to high organic loading (high food to microorganism ratio (F/M) conditions). Here, no flocs develop and biomass settling does not occur, resulting in a very turbid effluent from the clarifier. The correct remedial action for a dispersed growth problem is a reduction in the F/M of the system, usually done by raising the MLSS concentration. Dispersed growth problems often occur after a toxicity or hydraulic washout event when the activated sludge biomass is low and high F/M conditions prevail. Small, weak flocs can be formed in activated sludge that are easily sheared and subject to hydraulic surge flotation in the final clarifier leading to a turbid effluent. These small flocs, termed pin floc, consist only of floc-forming bacteria without a filament backbone and usually are <50um in diameter. Pin floc occurs most commonly at starvation conditions; a very low F/M and long sludge age. Chronic toxicity can also cause a pin floc condition. Free floating filaments can, at times, cause a dispersed growth problem. Here, the cause is filament-specific and is the same as for filamentous bulking.

Effluent From Settling Tanks

Outlet weirs or submerged orifices shall be designed to maintain velocities suitable for settling in the basin and to minimize short-circuiting. Weirs shall be adjustable, and at least equivalent in length to the perimeter of the tank. However, peripheral weirs are not acceptable as they tend to cause excessive short-circuiting. Large weir overflow rates result in excessive velocities at the outlet. These velocities extend backward into the settling zone, causing particles and flocs to be drawn into the outlet. Weir loadings are generally used up to 300 m^3/d/m. It may be necessary to provide special inboard weir designs as shown to lower the weir overflow rates. Clarified effluent is collected into a collection trough through an overflow weir, floating material or settled suspended solids should not be washed out with the clarified effluent, floating material retaining or entrapping baffle is provided, weir length is designed to avoid washout out of sludge Rectangular Primary Sedimentation Tanks. Weir channels are used for draining the effluent from the primary clarifier. Factor affecting the working of the component are; large weir loading a) re-suspension of particles settled near to effluent launders effluent weirs b) placed as far from the inlet as possible to increase weir length (i.e. to decrease weir loading) c) double-sided weirs can be used Typical weirs d) 90° V notch metal plates bolted onto the effluent collection through outlet zone. The weir may be placed at the opposite end of the rectangular basins a) through the entire width of tank b) through the length of the tank If the weir loading causes the required weir length to be greater than tank width the channel may be extended to a length of 1/3 the basin length (Reynolds) c) around the perimeter of centre – feed circular tanks d) at the centre of peripheral feed circular tanks.

SECONDARY TREATMENT: AEROBIC PROCESS

Suspended Growth Process

In the wastewater treatment processes when the microorganisms and bacteria treating the wastes or the growth of the microorganisms are in suspended form in the wastewater being treated using any of the aeration system, such as activated sludge processes (also aerated lagoons and aerobic digestion), the waste flows around and through the free-floating microorganisms, gathering into biological flocs that settle out of the wastewater. The settled flocs retain the microorganisms, meaning they can be recycled for further treatment. The various modes of the activated sludge process make use of suspended growth reactors. These reactors can be used for BOD (biochemical oxygen demand) removal, nitrification, and de-nitrification.

Activated Sludge

Sludge particles produced in wastewater by the growth of organisms in aeration tanks. The term 'activated' comes from the fact that the particles teem with bacteria, fungi, and protozoa. Activated sludge is different from primary sludge in that the sludge particles contain many living organ-isms that can feed on the incoming waste-water. The micro fauna present in the activated sludge of a wastewater treatment plant is mainly composed by bacteria, protozoa and metazoan. In the aeration tank of a wastewater treatment plant bacteria agglomerate generally in flocs. Attached to these flocs and swimming in the surrounding liquid one can find also bacteria, protozoa and metazoan may also be found (Da Motta et al., 2001). The activated sludge is the most widely used method to bring about stabilization in wastewater having organic matter constituents. The method depends on establishing and maintaining a population of degrading microorganisms and providing close contact of the degrading microorganisms and a supply of dissolved oxygen. The microorganisms feed and grow upon the oxidizable material in the wastewater and form a suspended floc of "activated sludge" in the water

Activated Sludge Terminology

The wastewater from the primary sedimentation tank, first enters with the mixture of returning activated sludge (RAS) from the secondary sedimentation

Figure 15. Schematic design of activated sludge plant

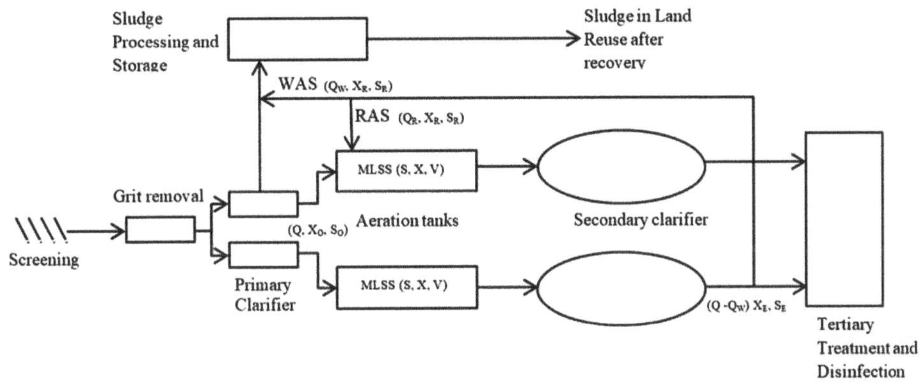

tank and the left over waste activated sludge (WAS) is goes for the sludge process and storage (biogas production or for land application as fertilizer). Effluent from the secondary sedimentation tank, the treated wastewater further goes for the disinfection process. The activated sludge process relies on the cultivation of a population of millions of microorganisms of many different types, mostly aerobic and facultative heterotrophic bacteria suspended in the wastewater, as it passes through a reactor (aeration tank). This suspension, referred to as mixed liquor (or Mixed Liquor Suspended Solids, MLSS), is supplied with oxygen and kept mixed by bubbling air through it. These are naturally occurring organisms; there is no need to supply them from an external source. As the organisms feed on the organic pollutants in the wastewater, the pollutants are converted to more organisms (biomass) and some by products.

Activated Sludge Process Equipment

A primary settler (or primary clarifier) may be introduced to remove part of the suspended solids present in the influent and this reduces the organic load to the activated sludge system. The biological reactor or aeration tank is filled with a mixture of activated sludge and influent, known as "mixed liquor". It is necessary to maintain certain mixed liquor suspended solid (MLSS) in the aerated tank maintain good removal efficiency. The aeration equipment transfers the oxygen necessary for the oxidation of organic material into the reactor, while simultaneously introducing enough turbulence to keep the sludge flocs in suspension. The continuous introduction of new influent results in a continuous discharge of mixed liquor to the secondary settler where separation of solids and liquid takes place. The liquid leaves the system as treated effluent, whereas some part of the sludge is re-circulated to the aeration tank called as 'return sludge' and rest of sludge is taken for anaerobic digestion.

Aeration Tank

The purpose of aeration tank is to dissolve oxygen into the wastewater in the aeration tank and to inter-mix the mixed liquor suspended solids in the aeration tank with the incoming wastewater. In the aeration tank of the wastewater treatment plants there are two methods of aeration: Mechanical and Diffused; each of these methods has its own system for delivery of air and mixing the mixed liquor with the wastewater. Mechanical: uses devices to either splash water into the air or mix air into the wastewater. Diffused: disperses compressed air to provide both oxygen to and the mixing of wastewater.

Suppose, Q is the flow rate of influent (m^3/d), QW is the flow rate of waste sludge (m^3/d), Q_r is the flow rate of return activated sludge (m^3/d), V is the volume of aeration tank (m^3), S0 is the influent soluble substrate concentration (BOD g/m^3), S is the effluent soluble substrate concentration (BOD g/m^3), X_o is the concentration of biomass in influent (g VSS/m^3), X_R is the concentration of biomass in return line from clarifier (g VSS/m^3), X_r is the concentration of biomass in sludge drains (g VSS/m^3) and X_e is the concentration of the biomass in the effluent (g VSS/m^3), VSS stands for volatile suspended solids

Design considerations of aeration tank are: -

1. Aeration period (4 – 6 hrs.)

$$t = \frac{Volume\left(m^3\right)}{Discharge\left(\frac{m^3}{day}\right)} \tag{31}$$

2. Organic loading rate (0.3 – 0.7 Kg/m^3)

$$OLR = \frac{Q \times Y_o\left(gm\right)}{Volume\left(m^3\right)} \tag{32}$$

Y_o = BOD in mg/L of the inflow sewage

3. Food/Mass ratio (0.2 – 0.4)

$$\frac{F}{M} = \frac{QY_o}{VX_t\left(MLSS\right)} \tag{33}$$

4. Sludge age, solid retention time or hydraulic retention time (5 – 8 days)

$$\theta_c = \frac{MLSS\ in\ the\ system}{MLSS\ leaving\ the\ system}$$

$$\theta_c = \frac{VX}{\left(Q - Q_W\right)X_e + Q_W X_W} = \frac{VX}{Q} \tag{34}$$

5. Sludge utilization rate

$$\left(u\right) = \frac{Q\left(S_o - S_e\right)}{V}\ \text{Kg/m}^3\text{/day} \tag{35}$$

6. Sludge produced

$$Q_W X_R = YQ\left(S_o - S_e\right) - K_d XV \tag{36}$$

Y = yield produced

Y = 1, for aerobic process TSS

$$\frac{1}{\theta_c} = Yu - K_d \tag{37}$$

7. Sludge volume index (50 – 150)

$$SVI = \frac{V_{observed}}{X_{observed}} = \frac{\frac{ml}{L}}{\frac{mg}{L}} = \frac{ml}{mg} \tag{38}$$

8. Size and volume of aeration tank

$$VX = \frac{YQ\left(S_o - S_e\right)\theta_c}{1 + K_d\theta_c} \tag{39}$$

$$\frac{1}{\theta_c} = \frac{Q\left(S_o - S_e\right)}{VX} \tag{40}$$

Example 11

Design a complete Sewage Treatment Plant of 1.5 MLD (average annual flow)

Basic data for the proposed STP

Average flow = 1.25 MLD= 1250 m^3/day

Peak flow = 3.125 MLD = 3125 m^3/day

G.L. near plant (av) R.L. 1860 m above MSL formation level

TWL in inlet chamber = 1863.50 m

TWL in outlet channel of FST = 1861.30

Sol. The proposed STP will enhance the following units

1. Inlet Chamber
2. Screen Chamber

Box 1. Sewage Characteristics

	Influent	Effluent
BOD, 20°C	295 mg/L	20 mg/L
Suspended solids	4653 mg/L	30 mg/L
pH	7.5	7.5

3. Grit Chamber
4. Aeration tanks
5. Secondary Clarifier
6. Sludge Drying Beds

Design of Inlet Chamber

TWL in inlet chamber = 1863.50

Peak flow = 3125 m^3/day = 0.038 m^3/sec

Plan dimension of inlet chamber = 1.3 X 1 m

Free Board = 10 m

Top of Inlet Chamber = RL 1864.50 m

Size of the bypass chamber by the side of inlet chamber = 1.3 X 1 m (minimum size)

The outflow from inlet chamber shall be taken to the screen chamber. A bye pass channel 0.60m X 1m or 400 mm Ø pipe shall be provided from bypass chamber up to final effluent channel to meet with any exigencies of the STP.

Design of Screen Chamber

2 nos screen chambers/channels shall be provided as per sound engineering practice. The flow from the inlet chamber to screen channels shall be controlled by C.I. penstock gates.

Q_{max} = 0.038 m^3/sec

Basic Assumptions:

Shape of Bar = M.S. Flats

Size = 10 mm X 50 mm (10 mm facing flow)

Clear spacing between the bars = 20 mm

Inclinations of bars with horizontal = 80° (cleaning manually)

Assuming velocity normal to screen = 0.8 m/sec

At peak flow net inclined area required = 0.038 / 0.8 = 0.047 m^2

Gross inclined area = 0.047 X 1.5 = 0.07 m^2

Gross vertical area required = 0.07 X Sin 80° = 0.069 m2

Provide submergence depth = 0.3 m

Width of channel = 0.069/0.3 = 0.23 m ; provide 0.3 m

$$\text{Check velocity in duct} = \frac{0.038}{0.30 \times 0.3} = 0.42 \text{ m/sec}$$

(Approach velocity u/s of screen) > 0.4 m/sec

Provide 20 bars of 10 mm x 50 mm @ 20 mm clear spacing

Screen chamber shall be 60 cm wide

U/s of screens, 2 nos C.I. penstock gates shall be provided (one for each channel)

Min. drop of 150 mm shall be provided in the bed of screen channel

Size of penstock gates: 2 nos of 350 x 450 mm size be provided

Design of Grit Chamber

Flow from screen channels shall be taken into Grit chamber, provided in duplicate. 2 no C.I. gates, one at each inlet and outlet, are provided with each grit chamber. One more C.I. gate shall be provided at the inlet to bye-pass channel in between two Grit Chambers.

Design Flow = 2.5 x 1.25 /2 = 1.56 mld = 1560 m^3/day

Surface loading = 1100 m^3/m^2/day

To account for turbulence and short circuiting reduce the surface loading to about 800 m^3/m^2/day

Area required = 1500/800 = 1.95 m^2

Provide 1.70 dia chamber or 1.7 x 1.7 m^2 chamber

Detention Time = 60 sec

$$\text{Volume} = \frac{1560 \times 60}{24 \times 3600} = 156/144 = 1.08\ m^3$$

Liquid depth = 1.08/1.95 = 0.554 m

Size of the Grit Chamber

= 1.70 m x 1.70 m (1.70 m dia) x 1.2 m (i.e. 0.554 + 0.6 F.B.) depth

Check for Horizontal Velocity

Cross sectional area of Grit chamber = 1.7 x 0.554 = 0.942 m^2

Velocity = $\dfrac{1560}{1.7 \times 0.554 \times 24 \times 3600}$ = 0.0192 m/sec = 1.9 cm/sec < 18 cm/sec. Hence O.K.

Grit generation = 0.05 m^3 per 1000 m^3 of sewage flow (assume)

Even though grit is continuously raked still 8 hours grit storage is provided for average flow

Storage Volume required = $\dfrac{1250 \times 8}{24} \times \dfrac{0.05}{1000}$ = 0.028 m^3

Grit storage area = $\dfrac{\pi}{4} \times 1.7^2 = 2.27$ m^2

Grit Storage depth = 0.028/2.27 = 0.012 m

Total liquid depth = 0.554 + 0.012 = 0.556; say 0.6 m

Provide grit chamber of size: 1.7 x 1.7 x x (0.60 + 0.60 FB) depth

Size = 1.7 m x 1.7m x 1.2m size

Outflow from grit chamber shall be carried out to aeration tanks through a 600 mm wide R.C.C. channel provided with fine bar screen (manually operated). The clear spacing between the bars shall be 10 mm.

No. of Tanks = 2

Average flow to each tank = 1.5 MLD/2 = 0.625 MLD = 625 m3/day = Q

The total BOD entering to STP = 295 mg/L

Assuming that negligible BOD is removed in screen and grit chamber,

The BOD of sewage coming to aeration = Yo = 295 mg/L

BOD left in effluent = Ye = 20 mg/L

Therefore BOD removed in activated plant = 295-20 = 275 mg/L

Minimum efficiency required in activated plant = 275/295 = 93%; O.K.

Since adopted extended aeration process can remove BOD up to 95-98%

Volume of aeration tank can be designed by assuming a suitable value of MLSS and θc (or F/M ratio)

Let us assume MLSS = 3000 mg/L (value is 3000 – 5000 mg/L)

F/M ratio = 0.12 (between 0.10-0.18)

Now using the above equation $\frac{F}{M} = \frac{Q}{V} \times \frac{Y_0}{X_t}$

where Q = 625 m^3/day

V=?

Yo = 295 mg/L

Xt = 3000 mg/L

F/M = 0.12 (assumed), therefore

$$0.12 = \frac{625}{V} \times \frac{295}{3000}$$

$$V = \frac{625}{0.12} \times \frac{295}{3000} = 512 \text{ m}^3$$

Aeration tank dimensions: let us adopt an aeration tank of liquid depth 3.5 m and 9 m width. Then,

The length of the tank = $= \frac{V}{B.D} = \frac{512}{9 \times 3.5} = 16.25$ m; say 16 m

Therefore, V provided = 16 m x 9m x 3.5m = 504 m^3

1. Check for aeration period or H.R.T. (t)

$$t = \frac{V}{Q} \times 24h = \frac{504}{625} \times 24 = 19.35 \text{ h} = 20 \text{ h}$$

O.K. (Lies between 10 -24 h)

2. Check for volumetric loading

Volumetric Loading $= \frac{QY_o}{V}$ gm of BOD5 /m^3

Vol. of tank $= \frac{625 \times 295}{504}$ = 366 gm/m^3 = 0.36 kg/m^3 O.K. (should be 0.2-0.4)

3. Check for return sludge ratio:

SVI should range from 50 – 150 ml/gm

$$\frac{Q_R}{Q} = \frac{X_t}{\frac{10^6}{SVI} - X_t}$$

Using SVI = 100 ml/gm; and Xt = 3000 mg/L, we have

$$\frac{Q_R}{Q} = \frac{3000}{\frac{10^6}{100} - 3000}$$

= 0.43; but it should be within 0.5 to 1.0

Hence we will provide 50% sludge recirculation giving SVI = 111, O.K.

4. Check for SRT (θc)

$$V.X_t = \frac{\alpha_y.Q\left(Y_0 - Y_E\right)\theta_c}{1 + K_e\theta_c}$$

α_y = 1.0 (constant for municipal sewage)
Ke = 0.06 d^{-1} (constant for a municipal sewage)
Y_o = 295 mg/L
Y_E = 20 mg/L

$V = 512\ m^3$
$Xt = 3000\ mg/L$
4. $Q = 625\ m^3/d$

$$504 \times 3000 = \frac{1.0 \times 625(295 - 20)\theta_c}{1 + 0.06\theta_c}$$

$$1 + 0.06\theta_c = 0.114\ \theta_c$$

$$0.054\ \theta_c = 1$$

$$\theta_c = \frac{1}{0.054} = 18.6 \text{ days ; O.K.}$$

Since it is between 10 -25 days

The adopted tank size is thus O.K. Hence adopt an aeration tank having an overall size of 16m x 9m x (3.5 + 0.6) m overall depth, with 0.6 m free board

The effluent from the aeration tank will be taken to the final (secondary) clarifier.

The inflow to the secondary clarifier shall be by means of 250 mm Ø C.I. pipe, which will give a velocity of 0.78 m/sec at peak flow.

Aerator Sizing:

BOD_5 applied to each tank = 295 mg/L

Average flow in each tank = 625 m^3/day

BOD5 to be removed in each tank = 625 x 0.295 = 184.375 kg/day

Say 184 kg/day = 8 kg/hr

Oxygen (O2) requirement = 1.2 kg/kg BOD applied

Peak oxygen demand = 125%

Oxygen transfer capacity of the aerator in standard conditions = 1.9 kg/KWH

= 1.41 kg/HP/hr

Oxygen transfer capacity of aerators at field conditions = 0.7 x 1.41 = 0.98 kg/HP/hr

Oxygen to be applied in each tank = 1.2 x 8 x 1.25 = 12 kg/hr

H.P. of aerators required = 12/0.98 = 12.24 H.P. = 14 H.P.

Provide two aerators each of 7 HP
Check for mixing considerations:
As per practice power required for mixing = 0.02 kW/m^3
Volume of each A.T. = 504m^3

HP required = 0.02 x 504 = 10.08 KW

Providing two aerators and considering gear efficiency as 97%

$$\text{H.P. of each aerators required} = \frac{10.08}{2 \times 0.97} = 5.2 \text{ HP}$$

Considering a power margin of 25% on motor rating
Motor of HP required = 5.2 x 1.25 = 6.50 ; say 7
Provide 2 no. of 7 HP motors/aerators in each tank
The dimension of the tank selected on the basis of suitable aerator motor is the zone of influence

The zone of influence for each aerator will be about 11 m2 with 3.5 m depth. The suitable aerator be selected and accordingly the depth and size of aeration tank adjusted. The submergence of the aerator will be between 75 mm to 115 mm.

Design of Secondary Clarifier

Secondary Clarifier
No. of secondary clarifier = 1 no
Average flow = 1250 m^3/day
Re-circulated flow say 50% = 625 m^3/day
Total inflow = 1875 m^3/day
Provide hydraulic detention time = 2 hrs

Volume of the tank (exclusive of hopper portion)=1875x2/24= 156.25 m^3

Assume liquid depth = 3.5 m

Area (superficial) = 156.25/3.5 = 44.64 m^2

Surface loading rate of average flow = 15 m^3/m^2/day

Surface area to be provided = 1250 /15 = 83m^2

(provide area greater of two i.e. 83m^2)

Dia of circular tank (d) = $d = \sqrt{\frac{83 \times 4}{\pi}}$ = 10.8 m; say 11 m

Actual area provided = 11x11 x π/4 = 95 m^2

Check for weir loading:
Average flow = 1250 m^3/day

Weir loading = $\frac{1250}{\pi \times 11}$ = 36.2 m^3/day/m O.K. less than 185 m^3/day/m

Provide a peripheral launder
Check for Solids Loading:
Recirculated flow = 625 m^3/day
Average flow = 1250 m^3/day
MLSS in the tank = 3000 mg/L

Total solids inflow = (1250 + 625) x 3 = 5625 kg/day

Solids loading = 5625/83 = 67.80 kg/day/m^2

Provide a clarifier of 11 m dia having liquid depth as 3.5 m
Hopper slope shall be 1 in 12
Free board will be 0.3 m

Sludge will be withdrawn from the clarifier through a C.I. pipe. The sludge will be taken to the return sludge pump house. The treated effluent from the secondary clarifier can be disposed of nearby water body or nallah

Return sludge pump house:

Total return flow = 625 m3/day = 0.434 m^3/min

Detention Time = 15 min

Volume of wet well = 0.435 x 15 = 6.5 m^3

Provide wet well = 2.5 x 1.5 x 1.8 m SWD

Provide dry well = 2.5 x 2.5 m

Size of annexe control room = 2.5 x 2.5 m

Provide two no. of pumps each of 0.625 MLD capacity in the dry well for returning the sludge to aeration tank

The return sludge pipe shall be 150 mm dia

Excess sludge shall be taken to the drying beds by providing necessary fittings in the main return sludge header. The sludge feed line to the sludge drying beds shall be 150 mm dia C.I. pipe.

Design of Sludge Drying Beds

Sludge applied to drying beds @ 100 kg/MLD

Sludge applied = 125 kg/day

Specific gravity = 1.015

Solid contents = 1.5%

Volume of the sludge = $= \dfrac{125}{1.5\%} \times \dfrac{1}{1000 \times 1.015}$ = 8.2 m^3/day

Considering monsoon etc. total no. of cycle in one year = 33

Period of each cycle = 365/33 = 11 days

Volume of sludge /cycle = 8.2 x 11 = 90.2 m^3

Spreading a layer of 0.3 m/cycle area of beds required = 90.2/0.3 = 300.67 m^2

Provide 4 beds of 1.2 m x 7 m

Thus providing = 336 m^2

Filtrate pump house and sump

Actual BOD_5 20°C removed per day = $1250 \times \dfrac{(295 - 20)}{1000}$ = 343.25 kg

Exccss wastcd sludge:

$$\theta_c = \frac{V.X_t}{Q_w.X_R}$$

$$18.6 = \frac{504 \times 3000}{Q_w.X_R}$$

$$Q_w.X_R = \frac{504 \times 3000}{18.6}$$

= 81.3 kg/day

Thus excess sludge produced = 81.3 kg/day

Assuming the excess sludge to contain 1% solids and specific gravity = 1.015, we have

Volumeofexcesssludge= $\frac{81.3}{1\% \times (1000 \times 1.015)} = \frac{81.3 \times 100}{1000 \times 1.015}$ =8m3/day=0.34 m^3/hr

Take detention time of 8 hours

Volume of we well = 8 x 0.34 = 2.64 m^3 for 1% concentration

Provide liquid depth = 1.0 m

Area required for 1% concentration of solids

= 2.64/1.0 = 2.64 m^2

Dia of wet well = $\sqrt{\frac{2.64 \times 4}{\pi}}$ = 1.83 m; provide 2.0 m Ø

Hence, provide 2 m Ø of the wet well for worst condition filtrate sump with 1.2 m liquid depth below the incoming pipe line.

This will be covered on top and a 4 m high pump room shall be constructed over it 2 no vertical non clog pumps each of 1 HP capacity shall be provided in this pump house. The filtrate from the pump house shall be pumped to the distribution chamber upstream of the aeration tank through 100 mm Ø C.I. pipe line.

The details of the designed units are shown in Fig.

Anaerobic Digester:

Primary clarifier efficiency in terms of

BOD = 30%

Solids = 65%

ASP efficiency = 90%

BOD load = 1250 m^3/day x 295 mg/L= 369 kg/day

SS load = 1250 m^3/day x 4653 mg/L = 5816 kg/day

Primary sludge solids = 0.65 x 5816 = 3780 kg/day

Primary effluent BOD for aeration tank = 0.7 x 369 = 258 kg/day

Assume $\theta_c = 8d$

BOD removal = 88%

Y = 0.5

Kd = 0.06 d^{-1}

S = 20 mg/L

Excess sludge = $= \dfrac{0.5 \times 0.88 \times 258}{1 + 0.06 \times 8}$ = 77 kg/d

Raw sludge composition:

- Primary solids:70% organics
    - Organic (kg/d): 57
    - Inorganic (kg/d): 24
- Secondary solids: 80% organics
    - Organic (kg/d): 62
    - Inorganic (kg/d): 15
- Total
    - Organic (kg/d): 119
    - Inorganic (kg/d): 39

Total solids = 158 kg/day
If there are 1% solids in sludge then

Total sludge (solid + water) = 158 x 100/1 = 15800 kg/day

Volume = 15800/1000 = 15.8 m3/d

In the tropical countries like India the temperature varies around 30° C where for the 80% conversion of BOD, detention time of 14 days is required

Volume of digester = 15.8 x 14 = 221 m^3

Size of the digester
Assume diameter = 7 m
Area = 38 m^2
Depth = 5.8 m $\approx$ 6 m
Calculation for Digested sludge and supernatant:
Anaerobic microbial mass synthesis
Assuming 6% solids in digester sludge
Kd = 0.03
Y = 0.1
θc = 14 days

$$\text{New mass synthesis} = \frac{Y\left(S_0 - S\right)}{1 + k_d\theta_c} = \frac{0.1 \times 0.88\left(119\right)}{1 + 0.03 \times 14} = 7.4 \text{ kg/d}$$

20% of organic will remain undigested = 0.20 X 119 = 23.8 kg/d

Digested sludge contain 6% solids and remaining 94% water

Total digested sludge = 23.8 + 7.4 + 39 = 70 kg/d

Therefore mass of sludge = 100 x total digested sludge /6 = 70 x 100/6 = 1167 kg/d

Volume = 1167 /1000 = 1.167 m3/d

Volume of supernatant = 15.8 – 1.167 = 14.633 m^3/d

Assumptions: In Biogas
Methane – 65%
CO_2 – 35%
1 kg of organic matter = 1 kg of COD
Organic Matter Input = 119 kg/d
Efficiency of digester = 80%
Organic matter metabolized = 95.2 kg/d = 96 kg/d
Anaerobic microbial mass synthesis = 7.4 kg/d = 7.5 kg/d

Organic matter converted into biogas = 96 – 7.5 = 88.5 kg/d

1 kg COD is equivalent to 350 L of methane

88.5 kg COD will produce = 88.5 x 350 = 30,975 L/d = 30.975 m^3/d = 31 m^3/d

Total biogas produced = 31/0.65 = 48 m^3/d

Aeration

The aeration tank are normally rectangular is shape (3 – 4 m deep, 4 – 6 m wide, and 20 – 200 m long), and the detention time 4 – 8 days for municipal sewage, air are continuously introduced into the tank for suspended growth process and aerobic treatment of wastewater. Methods uses for aeration are: - diffused aeration method, mechanical method, and combination of both.

1. **Diffused Aeration Method:** Compressed air are introduced under a pressure 35 – 70 KN/m^2 with the help of diffusers (porous plates and porous tubes) of quarts and crystalline aluminium, plates are generally square of 30*30 cm and 25 mm thickness fixed in the bottom of the tank. Tube diffusers are used of 600 cm long and 75 mm internal diameter and tubes thickness of 15 mm, working effective area 780 cm^2 and 1160 cm^2. Lots of air is wasted in this system, and may assumed that 4000 – 8000 m^3 of free air will be required per million litres of sewage, for BOD removal 100 m^3/day of air per of BOD.
2. **Mechanical Aeration Method:** Generally surface agitators and paddles are used in this method with aeration period 6 – 8 hrs. and the returning activated sludge should be

25-30 percent of the influent. For that two basic method can be used: Haworth system with 1 m deep, 70 m long and 1.5 m wide tank size, tank divided into no. of shells, Paddle revolved at the speed of 1.5 RPM, Detention time is 15 hrs., Returning sludge is of 15-20 percent of the influent and the Hartley system in this the arrangement of the blades are tilted in an angle

Settling Tank

The secondary sedimentation tanks or settling basins or clarifiers allow the microorganisms and other solids to settle after secondary or biological treatment. This sludge is pumped back into the inlet end of the primary sedimentation tanks and settles with the raw sludge. At this stage the BOD may be reduced to approximately 75 - 80% of the influent level and suspended solids are removed. Good design of the secondary sedimentation tank should provide 150 m^3/day/m of weir over flow rate. Solid loading rate should be 100 – 150 Kg/m^2/day and should not exceed the 250 Kg/m^2/day, the detention time should be of 1.5 – 2 hrs. The diameter to depth ratio of the circular tank is 5 and for the rectangular tank it is 7, depth of the tank should not exceed 3.5 – 4.5 m (circular)

Return Sludge

The sludge settled in the secondary clarifier which is returned to the aeration tank is known as returning activated sludge (RAS). The proper management of RAS is important and it provides a source of organisms that is returned to the activated sludge process. By changing the RAS rate, the operator can control the concentration of organisms in the aeration tank. Aerobic organisms will die for lack of oxygen if left in the clarifier too long. Caution has to be taken when increasing RAS rates as hydraulic overloading of the activated sludge system can occur and the increasing the RAS rates increases the volume in the aeration tank and in turn decreases the hydraulic detention time.

Waste Sludge

In the context of wastewater treatment residual is used to refer "sludge". The term sludge refers to the solids that are settled and separated during wastewater treatment. It is necessary to treat properly or dispose the sludge generated during the various stages of wastewater treatment like primary sedimentation, secondary sedimentation and sludge generated from advanced (tertiary) treatment, if any. The quantity of sludge generated depends upon the degree of treatment or quality of treated effluent required i.e., higher the degree of wastewater treatment, the larger the quantity of sludge to be treated and handled. Because of strict rules and regulations involving the handling and disposal of sludge, it has become necessary to reduce the volume of sludge in order to reduce the operating costs (approximately 50% of the plant cost) of treatment plants. Hence a properly designed and efficiently operated sludge processing and disposal system is essential to the overall success of the wastewater treatment plant. The sludge generated during the wastewater treatment can be classified into three categories:

Organic sludge originates from various sources such as PST, SST of ASP, SST of TF, etc. PST and SST sludge usually have 65% -75% volatile solids while sludge from TF of SST will have 45 to 70% volatile solids. The dry solids content of primary and secondary sludge after thickening is about 4 to 6% and specific gravity of 1.01. The separation of water is difficult from undigested sludge. Fuel value of undigested sludge is about 6500 to 8000 BTU/lb (15100 to 18600 KJ/kg of dry solids). After digestion sludge have volatile solids of about 32 to 48%, dry solids of 8 to 13%, and specific gravity of 1.03 to 1.05 and fuel value is about 8100 to 9300 KJ/kg of dry solids. Digested sludge is easy to dewater. Approximately 99.8% of coliform are destroyed from sewage sludge. Generally PST and SST sludge is thickened before digestion to reduce volume of sludge. Sludge treatment is an important aspect of wastewater treatment and account for 40 to 45% of the capital and operating costs

Overview of Activated Sludge Process

Activated sludge is a sludge particles reduced in wastewater by the growth of organisms in aeration tanks. The term 'activated' comes from the fact that the particles teem with bacteria, fungi, and protozoa and is different from primary sludge in the sense that the sludge particles contain many living organisms that can feed on the incoming wastewater. Described simply, screened wastewater is mixed with varying amounts of recycled liquid containing a high proportion of organisms taken from a secondary clarifying tank, and it becomes a product called mixed liquor. This mixture is stirred and injected with large quantities of air, to provide oxygen and keep solids in suspension. After a period of time, mixed liquor flows to a clarifier where it is allowed to settle. A portion of the bacteria is removed as it settles, and the partially cleaned water flows on for further treatment. The resulting settled solids, the activated sludge, are returned to the first tank to begin the process again. Today a number of variations of the basic process have been developed. This issue of Pipeline includes descriptions of three of the most common variations: Extended aeration, sequencing batch reactors, and oxidation ditches. The activated sludge plant is the most popular biological treatment process for larger installations.

A basic activated sludge process consists of several interrelated components: An aeration tank where the biological reactions occur, An aeration source that provides oxygen and mixing, A tank, known as the clarifier, where the solids settle and are separated from treated wastewater, A means of collecting the solids either to return them to the aeration tank, (return activated sludge [RAS]), or to remove them from the process (waste activated sludge [WAS]). Aerobic bacteria thrive as they travel through the aeration tank. They multiply rapidly with sufficient food and oxygen. By the time the waste reaches the end of the tank (between four to eight hours), the bacteria has used most of the organic matter to produce new cells. The organisms settle to the bottom of the clarifier tank, water. This sludge is pumped back to the aeration tank where it is mixed with the incoming wastewater or removed from the system as excess, a process called wasting. The relatively clear liquid above the sludge, the supernatant, is sent on for further treatment as required[12].

Factors Affecting Operation of the Activated Sludge Process

The most important component of any activated sludge plant is the biomass (microscopic bugs). In very simplified terms the bugs attach themselves to the waste, and digest the waste as a food supply, resulting in relatively clean water as the by-product. In an aeration basin there are several types of bugs that can be categorized into two main groups: carbonaceous (carbon eaters) and nitrogenous (chemical eaters like ammonia). Both groups have different duties, but as living organisms they all need oxygen to sustain life and reproduce. To understand the variations in oxygen demand, you need to first know how the bugs get into the aeration basin.

Effluent from the primary clarifier is mixed with the return activated sludge (RAS) from the secondary clarifier as it enters the aeration basin. This mixture, known as "mixed liquor," contains all the bugs the system needs to efficiently treat the wastewater. The hydraulic load, the amount of wastewater entering the aeration basin, determines the amount of RAS needed to maintain the proper bug population. With more wastewater entering the basin, more biological material and food exist for the bugs to eat, which increases their growth rate and demand for oxygen. After a few hours in the aeration basin (depending on the throughput of the plant), the mixed liquor is moved to the secondary clarifier. The mixed liquor settles there and a portion of the settled sludge is returned to the beginning of the aeration basin via RAS pumps. The excess sludge, known as waste activated sludge (WAS), is removed from the clarifier via WAS pumps and sent to the bio-solids stages of the plant for further treatment.

The greatest amount of oxygen is used near the intake portion of the aeration basin. In the secondary clarifier the bugs have been starved for oxygen because it was not added in this stage. Consequently, when the bugs return to the aeration basin, via the RAS line, they need to take a "deep breath." Also, the bugs use more oxygen when attaching themselves to the non settleable solids (dissolved and colloidal matter) upon reaching the aeration basin. As the flow passes through the basin, the demand for oxygen decreases. This is known as tapered aeration or step aeration. In a typical aeration basin, 40% of the oxygen demand occurs in the first quarter portion of the basin, 25% in the second quarter, 20% in the third, and 15% in last quarter portion. To obtain an "average" reading for monitoring and controlling the air addition, a dissolved oxygen sensor is normally installed about 2/3 of the length of the aeration basin from its intake. Some aeration plants use two dissolved oxygen sensors per basin to insure that there is enough oxygen in the first quarter of the basin to "wakeup" the bugs, without wasting energy by over-aerating the last quarter of the basin. Too much oxygen is not a problem for the bugs, but the cost to generate oxygen is one of the largest operating expenses of a wastewater treatment plant. If the supply of oxygen can be more tightly controlled, just enough to keep the bugs alive, then large amounts of money can be saved by generating only the amount of oxygen necessary. For this reason, multistage dissolved oxygen control is often used in large aeration basins. We have discussed how hydraulic loading, RAS control, and the type and quantity of bugs present can influence the oxygen demand in an aeration basin.

There are also several other factors that can influence proper control of the aeration basin: Temperature – All chemical reactions are affected by temperature. This is also true for waste consumption by the bugs, and the rate they multiply. When temperature increases, the rate at which the bugs eat and reproduce also increases. In fact, every 10°C increase doubles the bug reproduction rate, up to a limiting temperature. With an increase of bugs, the demand for oxygen increases dramatically. Most of the bugs achieve optimum efficiency between 10° and 40°C. pH – As with temperature, the pH has a considerable influence on the ability of the bugs to work efficiently. When the pH is too low or too high, the bugs lose the ability to convert the food to energy and raw materials. The aeration basin pH for most systems should be kept between 6.5 pH and 8.5 pH. Toxicity– Some organic materials and heavy metals are toxic to the bugs, reducing their efficiency or even destroying them. If the bugs get sick or the weaker strains die, the micro--universe suffers, reducing the plant efficiency and increasing the potential of untreated waste.

Growth Curve

Bacteria can reproduce by binary fission, either by sexual mode or by budding. By binary fission, one organism becomes two in a given generation time which would vary from species to species. Some species take less than 20 min while others may take days together to multiply. The general growth pattern of bacteria in a batch reactor is characterized by four identifiable distinct growth phases.

The curve shown in Fig 15 represents what occurs in a batch reactor in which, at time zero, the substrate and nutrients are present in excess and only a very small population of biomass exists.

Figure 16. Schematic diagram of microbial growth curve.

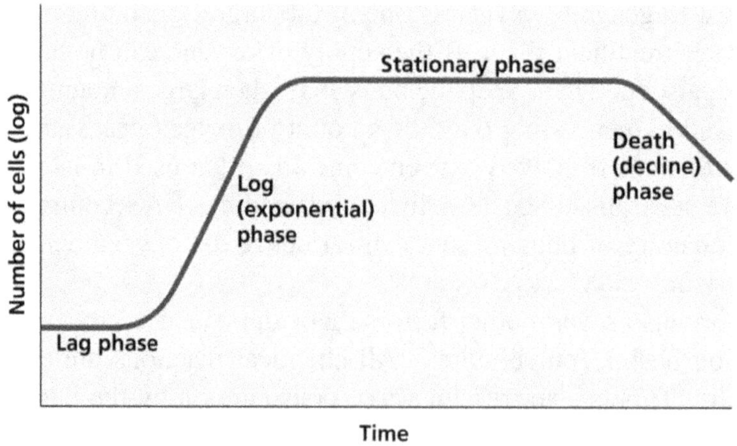

1. **The Lag Phase:** Upon addition of the biomass, the lag phase represents the time required for the organisms to acclimate to their new environment before significant cell division and biomass production occurs. During the lag phase enzyme induction may be occurring and /or the cells may be acclimating to changes in salinity, pH, temperature etc.
2. **The Exponential Growth Phase:** During the exponential growth phase, bacterial cells multiply at their maximum rate, as there is no limitation due to substrate or nutrients. The biomass growth curve increases exponentially during this period. With unlimited substrate and nutrients the only factor that affects the rate of exponential growth is temperature.
3. **The Stationary Phase:** During this phase, the biomass concentration remains relatively constant with time. In this phase bacterial growth is no longer exponential and the rate of growth is reduced by the death of cells.
4. **The Death Phase:** In the death phase, the substrate has been depleted so that, no growth occurs and the change in biomass concentration is due to cell death. An exponential decline in the biomass concentration is often observed, as an approximate constant fraction of the biomass remaining is lost each day.

Activated Sludge Formation

Activated sludge is also the name given to the active biological material produced by activated sludge plants. Excess sludge is called "surplus activated sludge" or "waste activated sludge" and is removed from the treatment process to keep the ratio of biomass to food supplied in the wastewater in balance. This sewage sludge is usually mixed with primary sludge from the primary clarifiers and undergoes further sludge treatment for example by anaerobic digestion, followed by thickening, dewatering, and composting and land application. The amount of sewage sludge produced from the activated sludge process is directly proportional to the amount of wastewater treated. The total sludge production consists of the sum of primary sludge from the primary sedimentation tanks as well as waste activated sludge from the bioreactors. The activated sludge process produces about 70–100 kg/ML of waste activated sludge (that is kg of dry solids produced per ML of wastewater treated; one mega litre (ML) is 10^3 m^3). A value of 80 kg/ML is regarded as being typical. In addition, about 110–170 kg/ML of primary sludge is produced in the primary sedimentation tanks which most - but not all - of the activated sludge process configurations use. A variant of the activated sludge process is the Nereda process where aerobic granular sludge is developed by applying specific process conditions that favour slow growing organisms.

Activated Sludge Performance-Controlling Factors

The general process control method is to monitor sludge blanket level, SVI (Sludge Volume Index), MCRT (Mean Cell Residence Time), F/M (Food to Microorganism), as well as the biota of the activated sludge and the major nutrients DO (Dissolved oxygen), nitrogen, phosphate, BOD (Biochemical oxygen demand), and COD (Chemical oxygen demand). In the reactor/aerator and clarifier system, the sludge blanket is measured from the bottom of the clarifier to the level of settled solids in the clarifier's water column; this, in large plants, can be done up to three times a day. The SVI is the volume of settled sludge in millilitres occupied by 1 gram of dry sludge solids after 30 minutes of settling in a 1000 millilitre graduated cylinder. The MCRT is the total mass (lbs) of mixed liquor suspended solids in the aerator and clarifier divided by the mass flow rate (lbs/day) of mixed liquor suspended solids leaving as WAS and final effluent. The F/M is the ratio of food fed to the microorganisms each day to the mass of microorganisms held under aeration. Specifically, it is the amount of BOD fed to the aerator (lbs/day) divided by the amount (lbs) of Mixed Liquor Volatile Suspended Solids (MLVSS) under aeration. Note: Some references use MLSS (Mixed Liquor Suspended Solids) for expedience, but MLVSS is considered more accurate for the measure of microorganisms. Again, due to expedience, COD is generally used, in lieu of BOD, as BOD takes five days for results. Based on these control methods, the amount of settled solids in the mixed liquor can be varied by wasting activated sludge (WAS) or returning activated sludge (RAS) [13].

Alkalinity

The alkalinity of water is a measure of its capacity to neutralize acids. It also refers to the buffering capacity, or the capacity to resist a change in pH. For wastewater operations, alkalinity is measured and reported in terms of equivalent calcium carbonate (CaCO3). Alkalinity is commonly measured to a certain pH. For wastewater, the measurement is total alkalinity, which is measured to a pH of 4.5 SU. Even though pH and alkalinity are related, there are distinct differences between these two parameters and how they can affect your facility operations.

One of the primary pollutants in municipal and food processing wastewaters is ammonia. The nitrification process is used to convert the ammonia in the wastewater to nitrate. Nitrification is a two-step biological process utilizing two species of nitrogen converting bacteria. These species of bacteria are most active in the pH range of 7 to 8. During the nitrification process, hydrogen ions are released and alkalinity is consumed as the acid is neutralized. For every 1.0 mg of ammonia converted to nitrate, 7.14 milligrams (mg) $CaCO_3$equivalent are consumed. The generation of acid during ammonia conversion, and the need to maintain the proper pH, is the reason why an alkali is added to the system. Nitrification of the wastewater is the single largest factor which leads to the consumption of alkalinity and the need to add alkali to the treatment system.

How much alkali is added to the system is dependent on a number of interrelated factors. The amount of alkali added is determined by the amount of pollutants in the incoming waste, the type of treatment that is used in the plant, the amount of natural alkalinity in the influent water, the pH of the influent waste stream, the permitted pH of the effluent discharged from the plant, the number of gallons of waste processed by the plant, and whether the plant denitrifies the effluent prior to final treatment and discharge.

Calcium and magnesium are essential elements bacteria need to carry on metabolic functions and reproduce. Other essential elements include nitrogen, phosphorus, potassium, iron, sulphur, oxygen, carbon and hydrogen. Without these elements present, the bacteria in treatment plants would not function properly, and could result in an over-abundance of nuisance bacteria — the kinds that cause odors or inhibit settling.

We need some alkalinity to buffer against drops in pH values. Processes that biologically produce acids or acidic chemicals include:

1. Biological nitrification (the conversion of ammonium to nitrite then nitrate)
2. Anaerobic conditions in sewer systems
3. Anaerobic conditions in primary clarifiers
4. Anaerobic sludge digestion processes
5. Anaerobic fermentation basins in biological phosphorus removal systems
6. Chemical coagulant addition (aluminium sulphate, ferric sulphate, ferric chloride)
7. Pure gaseous chlorine for disinfection

When these biological conditions occur in a treatment plant, or when acidic chemicals are added, the free hydrogen (H^+) in the acid reacts with the negatively charged alkalinity, and the two effectively neutralize each other. If the water contained only the exact amount of alkalinity required to neutralize the acids, there would not be enough alkalinity remaining to protect the final pH value from falling if any additional acid were formed or added downstream.

Biological processes like nitrification and anaerobic digestion rely on alkalinity. Without alkalinity, organic acids formed during these processes would drive the pH down to a point where the bacteria would be inhibited or could no longer survive.

For instance, during the acid-formation stage of anaerobic digestion, volatile fatty acids are produced as acid-forming bacteria feed on the viscous, nearly septic sludge. A second group of bacteria, methanogens, then consume the volatile fatty acids. From this reaction, methanogens produce methane and bicarbonate alkalinity. The alkalinity they produce helps buffer the acid produced by the volatile acid formers.

In a properly operated anaerobic digester, the ratio of volatile acid to alkalinity is between 0.1 to 0.25 parts acid for every one part alkalinity per litre. If a digester is overfed and volatile acids are rapidly increasing, the methane formers can't consume the acids fast enough. This causes the alkalinity to become depleted (R., & C., 2018).

For example, if the volatile acid climbs rapidly to 1,500 mg/L and the alkalinity is steady at 3,000 mg/L as CaCO3, then the ratio becomes 0.5:1, and methane production slows down or even stops. If the operator does not take corrective action, the digester may become sour and stop working completely. In fact, if the operator did not pay attention to alkalinity and used pH as the sole operating process control parameter, the digester could become sour before the pH finally indicated an operating problem.

Dissolved Oxygen

Wastewater aeration is the process of adding air into wastewater to allow aerobic bio-degradation of the pollutant components. It is an integral part of most biological wastewater treatment systems. Unlike chemical treatment which uses chemicals to react and stabilize contaminants in the wastewater stream, biological treatment uses microorganisms that occur naturally in wastewater to degrade wastewater contaminants.

Aeration provides oxygen to bacteria for treating and stabilizing the wastewater. Oxygen is needed by the bacteria to allow biodegradation to occur. The supplied oxygen is utilised by bacteria in the wastewater to break down the organic matter containing carbon to form carbon dioxide and water. Without the presence of sufficient oxygen, bacteria are not able to biodegrade the incoming organic matter in a reasonable time. In the absence of dissolved oxygen, degradation must occur under septic conditions which are slow, odorous, and yield incomplete conversions of pollutants. Under septic conditions, some of the biological process converts hydrogen and sulphur to form hydrogen sulphide and transform carbon into methane. Other carbon will be converted to organic acids that create low pH conditions in the basin and make the water more difficult to treat and promote odour formation. Biodegradation of organic matter in the absence of oxygen is a very slow biological process.

Oxygen is required in the activated sludge process for the oxidation of a part of the influent organic matter and also for the endogenous respiration of the micro-organisms in the system. The total oxygen requirement of the process may be formulated as follows:

$$O_2 Required\left(\frac{g}{d}\right) = \frac{Q\left(S_o - S_e\right)}{f} - 1.42 Q_W X_R \qquad (41)$$

Where, f = ratio of BOD_5 to ultimate BOD and 1.42 = oxygen demand of biomass (g/g). The formula does not allow for nitrification but allows only for carbonaceous BOD removal.

pH

The hydrogen ion concentration of the culture medium has a direct influence on microbial growth. Most biological treatment systems operate best in a neutral environment. The general range for operation of activated – sludge systems is between pH 6.5 and 8.5. At pH 9.0 and above, microbial activity is inhibited. Below 6.5, fungi are favoured over bacteria in the competition for food. The methane-forming bacteria in anaerobic digestion have a much smaller pH tolerance range. General limits for anaerobic digestion are pH 6.7-7.4 with optimum operation a pH 7.0-7.1.

Fungi, for example, do well in an acidic environment. Most bacteria and protozoa, however, grow best in neutral (pH 7) environments. Abnormal or irregular pH in biological treatment processes can result in a significant decrease in the rate of removal of organic compounds from the environment, which will affect the biochemical oxygen demand (BOD) measurements.

Many plants must control the pH of their process effluent within an acceptable range before it is mixed with biomass in the Aerated Stabilization Basin (ASB) or Activated Sludge System. Even short-term exposure (exposure lasting less than one minute) to extreme pH causes significant microbial destruction. Some influents (process effluents) are slightly on the basic side with pH between 9.0 and 10.5. Because bacteria generate CO_2 (an acidic gas) as a by-product of metabolism, they will self-regulate the pH to some extent, as long as the pH is not so severe as to completely stop the bacteria's metabolism.

pH approximates the concentration of hydrogen ions in a solution. The pH value is the negative logarithm (base 10) of the concentration of H+ ions in the solution. In the laboratory, pH is measured by electrometric pH measurement which is the determination of the activity of the hydrogen ions by potentiometric measurement using a standard hydrogen electrode and a reference electrode. The pH probe is placed in the sample (while stirring) and the number is recorded once the readings have stabilized. Because it is on a logarithmic scale, it will take ten times as much acid or base (caustic) to raise or lower the pH two units as it did to raise or lower it one unit. For example, if it takes 10 gallons of a particular acid to lower the pH of an influent from 10 to a pH of 9, it will take 100 gallons to lower it from a pH of 10 to a pH of 8.

Mixed Liquor Suspended Solids, Mixed Liquor Volatile Suspended Solids, and Mixed Liquor Total Suspended Solids

Mixed liquor is a combination of sludge and water removed from the clarifier in the wastewater treatment process and reintroduced into an earlier phase of the treatment process. The mixed liquor contains microorganisms which digest the wastes in the raw water. This lab will show you how to test for MLSS and MLVSS in the mixed liquor. Mixed Liquor Suspended Solids (MLSS) is a test for the total suspended solids in a sample of mixed liquor. This test is essentially the same as the test you performed for TSS in the last lab, except for the use of mixed liquor as the water sample. In addition, the concentration of suspended solids found in the mixed liquor is typically much greater than that found in the raw or treated water. MLSS concentrations are often greater than 1,000 mg/L, but should not exceed 4,000 mg/L.

MLVSS, or Mixed Liquor Volatile Suspended Solids, is a test for the amount of volatile suspended solids found in a sample of mixed liquor. Volatile solids are those solids which are burnt up when a sample is heated to 550°C. Most of the volatile solids in a sample of mixed liquor will consist of microorganisms and organic matter. As a result, the volatile solids concentration of mixed liquor is approximately equal to the amount of microorganisms in the water and can be used to determine whether there are enough microorganisms present to digest the sludge. In a wastewater treatment plant, operators should test for MLSS three times per week and for MLVSS once per week. Both tests should use grab samples taken from the same location in the treatment plant.

Return Activated Sludge Rate and Concentration

Settled activated sludge that is collected in the secondary clarifier or the membrane basin and returned to the aeration basin to mix with incoming raw or primary settled wastewater, from the primary sedimentation tank. The returning sludge should be mixed in 25 – 50 percent of the influent in the activated sludge treatment process and can be calculated as: -

$$\frac{Q_R}{Q} = \frac{Returning\ sludg\ rate\left(\frac{m^3}{d}\right)}{Influent\ rate\ of\ sewage\left(\frac{m^3}{d}\right)} \tag{42}$$

$$\frac{Q_R}{Q} = \frac{X_t}{X_R - X_t} \tag{43}$$

$$X_t = MLSS\ in\ \frac{mg}{L} \tag{44}$$

$$X_R = Return\,activated\,sludge\,in\,\frac{mg}{L},= \frac{10^6}{SVI} \tag{45}$$

Temperature

Wastewater temperatures normally range between 10 and 20OC. In general, the temperature of the wastewater will be higher than that of the water supply. This is because of the addition of warm water from households and heating within the plumbing system of the structure. Depending on the geographic location, the mean annual temperature of wastewater varies. The temperature of water is a very important parameter because of its effect on chemical reactions and reaction rates, aquatic life, and the suitability of the water for beneficial uses. Increased temperature, for example, can cause a change in the species of fish that can exist in the receiving water body.

In addition, oxygen is less soluble in warm water than in cold water. The increase in the rate of biochemical reactions that accompanies an increase in temperature, combined with the decrease in the quantity of oxygen present in surface waters, can often cause serious depletion in dissolved oxygen concentrations in the summer months. Optimum temperatures for bacterial activity are in the ranges from about 25 to 35 ℃. Aerobic digestion and nitrification stop when the temperature rises to 50℃. When the temperature drops to about 15℃, methane-producing bacteria become quite inactive, and at about 50℃, the autotrophic-nitrifying bacteria practically cease functioning.

Increase in deoxygenating coefficient with the increase of incubation temperature of the sample, and obtained the maximum degradation rate at 30 ℃. Considering that, with the increase of incubation temperature T, there is an increase in the value of K, it is possible then to reduce the incubation time to achieve a value similar to the one that would be obtained if the incubation occurred at 20 ℃. The relation between temperature and

Table 7. Temperature classification of biological process

Type	Temperature Range, °C	Optimum Temperature Range, °C	Temperature Range, °F	Optimum Temperature Range, °F
Psychrophilic	10 to 30	12 to 18	50 to 86	53.6 to 64.4
Mesophilic	20 to 50	25 to 40	68 to 122	77 to 104
Thermophilic	35 to 75	55 to 65	95 to 167	131 to 149

deoxygenating coefficient was demonstrated by the Arrhenius relation modified by STREETER & PHELPS (1925)

$$K_T = K_{20} \left(\theta\right)^{(T-20)}$$

$$K_T = K_{20} \left(1.047\right)^{(T-20)}$$

Where, K = deoxygenating coefficient, T = temperature

Sludge Blanket Depth

Sludge blanket depth is an operational process control test in which the depth of sludge accumulated at the bottom of a secondary clarifier is measured; most commonly accomplished using a "sludge judge." Prudent practice involves monitoring sludge blanket depth a minimum of once per day or once per shift for multi shift operations. A sludge blanket depth of target of 0.3 to 0.6 m (1 to 2 ft) is adequate for most treatment facilities. For best performance, a good rule-of-thumb is to maintain the blanket of 0.6 m (2 ft) or less at all times.

Sludge blanket depth in the clarifier should be measured at the same time each day. The best time is during the period of maximum daily flow, because the clarifier is operating under the highest solids loading rate. Adjustments in the RAS flow rate should be needed only occasionally if the activated sludge process is operating properly and the anaerobic treatment system is running smoothly.

An additional advantage of monitoring the sludge blanket depth is that problems, such as improperly operating sludge collection equipment, will be observed due to irregularities in the blanket depth. A plugged pick-up on a clarifier sludge collection system would cause sludge depth to increase in the area of the pick-up, and decrease in the areas where the properly operating pick-ups are located. These irregularities in sludge blanket depth are easily monitored by measuring profiles of blanket depth across the clarifier.

Holding solids too long generally causes not only septic conditions, which can generates odors, but also generates low DO conditions which can generate specific compounds with sulphides and/or organic acids. Limited nutrients are available also to the bacteria in the clarifier. All of these conditions promote the growth of filamentous bacteria. Holding times of influent should always be kept to a minimum. Generation of septic conditions can also contribute to ashing in the clarifier, since the solids are turning anaerobic in the bottom of the clarifier and generate gases that cause clumps of solids to float to the top. This increases solids carryover. Polymer consumption goes up, solids handling is increased. Since many plants have a limitation on sludge dewatering, either due to draw down on the clarifier, pumping capabilities or mechanical sludge dewatering abilities, these variables are critical.

Types of Ponds: Oxidation Pond and Stabilisation Ponds

Stabilization Ponds

1. The stabilization ponds are open flow through basins specifically designed and constructed to treat sewage and biodegradable industrial wastes. They provide long detention periods extending from a few to several days. Waste stabilization ponds (WSPs) are increasingly applied for domestic wastewater treatment, and offers the completely natural purifying processes (Ho, Van Echelpoel, & Goethals, 2017).
2. The only energy they use is direct solar energy, so they do not need any electromechanical equipment, saving expenditure on electricity and more skilled operation. They do require much more land than conventional electromechanical treatment processes such as activated sludge – but land is an asset which increases in value with time, whereas money spent on electricity for the operation of electromechanical systems is gone forever. WSPs are one of the main natural wastewater treatment methods. They are man-made earthen basins, comprising at any one location one or more series of anaerobic, facultative and, depending on the effluent quality required, maturation ponds. WSPs are particularly suited to tropical and subtropical countries since sunlight and ambient temperature are key factors in their process performance (Waste Stabilization Ponds, 2004).
3. Pond systems, in which oxygen is provided through mechanical aeration rather than algal photosynthesis, are called aerated lagoons.
4. Stabilization ponds may be aerobic, anaerobic or facultative.
5. Aerobic ponds, also known as high-rate algal ponds, can maintain dissolved oxygen throughout the 30–45 cm-deep pond because of algal photosynthetic activity (USEPA 2011). Aerobic ponds are shallow ponds with depth less than 0.5 m and BOD loading of 40-120 kg/ha.d so as to maximize penetration of light throughout the liquid depth. Such ponds develop intense algal growth. Photosynthetic activity supplies oxygen

during the day, while at night the wind creates aeration due to the shallow depth of the pond (Davis and Cornwell 2008).

6. Anaerobic ponds operate without the presence of dissolved oxygen. Under methanogenic conditions, the major products are carbon dioxide and methane (Quiroga 2011). Anaerobic ponds are used as pre-treatment of high strength wastes with BOD load of 400-3000 kg/ha.d Such ponds are constructed with a depth of 2.5-5m as light penetration is unimportant. The driving force behind treatment is sedimentation. Helminths settle to the bottom of the pond, and bacteria and viruses are removed by attaching to settling solids within the pond or die with the loss of available food or by the presence of predators. In practice, anaerobic ponds are usually incorporated alongside facultative ponds (Martinez et al. 2014).
7. A facultative pond is a treatment unit with anaerobic and aerobic conditions. A typical pond is divided into an aerobic surface region consisting of bacteria and algae, an anaerobic bottom region, consisting of anaerobic bacteria, and a region in between anaerobic and aerobic conditions where bacteria can thrive in both conditions (Kayombo et al. 2010). Facultative pond functions aerobically at the surface while anaerobic conditions prevail at the bottom. They are often about 1 to 2 m in depth. The aerobic layer acts as a good check against odour evolution from the pond.

Mechanism of Purification

The functioning of a facultative stabilization pond and symbiotic relationship in the pond are shown below. Sewage organics are stabilized by both aerobic and anaerobic reactions. In the top aerobic layer, where oxygen is supplied through algal photosynthesis, the non-settleable and dissolved organic matter is oxidized to CO2 and water. In addition, some of the end products of partial anaerobic decomposition such as volatile acids and alcohols, which may permeate to upper layers, are also oxidized periodically. The settled sludge mass originating from raw waste and microbial synthesis in the aerobic layer and dissolved and suspended organics in the lower layers undergo stabilization through conversion to methane which escapes the pond in form of bubbles.

Various factors affect pond design:

1. Wastewater characteristics and fluctuations.
2. Environmental factors (solar radiation, light, temperature)
3. Bacterial growth patterns and decay rates.

4. Solids settlement, gasification, upward diffusion, sludge accumulation.

The depth of aerobic layer in a facultative pond is a function of solar radiation, waste characteristics, loading and temperature. As the organic loading is increased, oxygen production by algae falls short of the oxygen requirement and the depth of aerobic layer decreases. Further, there is a decrease in the photosynthetic activity of algae because of greater turbidity and inhibitory effect of higher concentration of organic matter.

Gasification of organic matter to methane is carried out in distinct steps of acid production by acid forming bacteria and acid utilization by methane bacteria. If the second step does not proceed satisfactorily, there is an accumulation of organic acids resulting in decrease of pH which would result in complete inhibition of methane bacteria. Two possible reasons for imbalance between activities of methane bacteria are: (1) the waste may contain inhibitory substances which would retard the activity of methane bacteria and not affect the activity of acid producers to the same extent. (2) The activity of methane bacteria decreases much more rapidly with fall in temperature as compared to the acid formers. Thus, year round warm temperature and sunshine provide an ideal environment for operation of facultative ponds.

Algal Growth and Oxygen Production

Algal growth converts solar energy to chemical energy in the organic form. Empirical studies have shown that generally about 6% of visible light energy can be converted to algal energy. The chemical energy contained in an algal cell averages 6000 calories per gram of algae. Depending on the sky clearance factor for an area, the average visible radiation received can be estimated as follows:

Avg. radiation = Min. radiation + [(Max. radiation - Min. radiation) x sky clearance factor]

Oxygen production occurs concurrently with algal production in accordance with following equation:

$$106CO_2 + 16NO_3 + HPO_4 + 122H_2O + 18H \rightarrow C_{106}H_{263}O_{110}N_{16}P + 138O_2$$

On weight basis, the oxygen production is 1.3 times the algal production.

Areal Organic Loading

The permissible areal organic loading for the pond expressed as kg BOD/ha.d will depend on the minimum incidence of sunlight that can be expected at a location and also on the percentage of influent BOD that would have to be satisfied aerobically. The Bureau of Indian Standards has related the permissible loading to the latitude of the pond location to aerobically stabilize the organic matter and keep the pond odour free. The values are applicable to towns at sea levels and where sky is clear for nearly 75% of the days in a year. The values may be modified for elevations above sea level by dividing by a factor (1 + 0.003 EL) where EL is the elevation of the pond site above MSL in hundred meters.

Detention Time

The flow of sewage can approximate either plug flow or complete mixing or dispersed flow. If BOD exertion is described by first order reaction, the pond efficiency is given by:

For plug flow: $L_e/L_i = e^{-kt}$

For complete mixing: $$\frac{L_e}{L_i} = \frac{1}{1+kt} \quad (46)$$

For dispersed flow the efficiency of treatment for different degrees of intermixing is characterized by dispersion numbers. Choice of a larger value for dispersion number or assumption of complete mixing would give a conservative design and is recommended.

Depth

Having determined the surface area and detention capacity, it becomes necessary to consider the depth of the pond only in regard to its limiting value. The optimum range of depth for facultative ponds is 1.0 - 1.5 m.

Aquatic Plant Systems

Aquatic systems in waste treatment are either free floating growths harnessed in the form of built-up ponds for waste treatment such as duckweed and hyacinth ponds or rooted vegetation's (reeds) which emerge out of shallow waters cultivated in constructed wetlands.

1. Natural wetlands exist all over the world. They generally have saturated soil conditions and abound in rooted vegetation which emerges out of shallow waters in the euphotic zone. They may also have phytoplankton. Natural wetlands can be integrated with wastewater treatment systems.
2. Constructed wetlands are man-made for treatment of wastewater, mine drainage, storm drainage, etc. They have rooted vegetation.

Aquatic plant ponds consisting of free floating macrophytes, such as water hyacinths, duckweeds, etc. have been cultured in ponds either for their ability to remove heavy metals, phenols, nutrients, etc. from wastewaters or to assist in giving further treatment to pre-treated wastewaters to meet stringent discharge standards while at the same time producing new plant growths for their gas production or food value.

Aerated Lagoon

- **Sewage Lagoon:** consists of a large, shallow earthen basin in which wastewater is retained long enough for natural purification processes. Classification of lagoons is based on degree of mechanical mixing provided.
- **Aerobic Lagoon:** The reactor is called an aerobic lagoon, when sufficient energy is supplied to keep the entire contents, including the sewage solids, mixed and aerated. To meet suspended-solids effluent standards, solids are removed from the effluent coming from an aerobic lagoon.
- **Facultative Lagoon:** In facultative lagoon, only enough energy is supplied to mix the liquid portion of the lagoon, solids settle to the bottom in areas of low velocity gradients and proceed to degrade anaerobically and this process is different from facultative pond only in the method by which oxygen is supplied. Facultative lagoons are assumed to be completely mixed reactors without biomass recycle.
- **Aerobic Lagoons With Solid Recycle:** The aerobic lagoon with solids recycle is same as extended aeration activated-sludge process, but an earthen (typically lined) basin is used in place of a reinforced-concrete reactor basin. It is necessary that the aeration requirement for

an aerobic lagoon with recycle must be higher than the values for an aerobic flow-through lagoon to maintain the solids in suspension.

Example 12

Wastewater flow from a small community averages 3400 m^3/d during the winter and 6600 m^3/d during the summer. The average temperature of the coldest month is 10°C, and the average temperature of the warmest month is 30°C, the average BOD_5 is 200 mg/L 70% being soluble. The reaction coefficient k is 0.23 d-1 at 20°C, and the value of temperature coefficient is 1.06. Prepare a preliminary design for a facultative pond treatment system for the community to remove 90% of the soluble BOD.

1. Find volume of facultative lagoon to remove 90% of the soluble of BOD.
2. Find the dimensions of three square lagoons in series with depth 1.5 m.

Sol. Given:

1. Rate constant at given temperature
2. Summer: $k_{25} = 0.23(1.06)^{30\text{-}20} = 0.411\ d^{-1}$
3. Winter: $k_{10} = 0.23(1.06)^{10\text{-}20} = 0.128\ d^{-1}$
4. Volume of lagoon
5. Summer:

$$\frac{S}{S_0} = \frac{1}{1 + K\left(V / Q\right)} \tag{47}$$

$$\frac{20}{200} = \frac{1}{1 + 0.411\left(\dfrac{V}{6600}\right)}$$

6. $V = 144525.5\ m^3$
7. Winter: $\dfrac{20}{200} = \dfrac{1}{1 + 0.128\left(V / 3400\right)}$
8. $V = 239062\ m^3$
9. Dimensions of three square lagoons in series

10. See Eq. 48:

$$\frac{S_n}{S_0} = \frac{1}{\left(1 + k\left(\frac{V_i}{n \times Q}\right)\right)^n} \tag{48}$$

$$\frac{200}{20} = \left(1 + \frac{0.411 \times V_i}{3 \times 6600}\right)^3$$

11. $V_i = 55607.13\ m^3$
12. Winter: $\frac{200}{20} = \left(1 + \frac{0.128 \times V_i}{3 \times 3400}\right)^3$
13. $V_i = 91980.8\ m^3$

Attached Growth Process

It is the wastewater treatment process in which the microorganisms and bacteria treating the wastes are attached to the media in the reactor. The wastes being treated flow over the media. Trickling filters and rotating biological contactors are attached growth reactors. These reactors can be used for BOD removal, nitrification, and denitrification.

Trickling Filters

A trickling filter, also called trickling bio-filter, bio-filter, biological filter and biological trickling filter, is a fixed-bed, biological reactor that operates under (mostly) aerobic conditions. Pre-settled wastewater is continuously 'trickled' or sprayed over the filter. As the water migrates through the pores of the filter, organics are aerobically degraded by the biofilm covering the filter material.

Trickling Filter Definitions

Trickling filter commercially established cost effective technique for biodegradable loads and ammonia removal from municipal and industrial effluent, adopt for the secondary treatment of sewage. The main function of trickling filter is to reduce soluble BOD_5 in the wastewater, but it hardly achieves value beyond 85% in terms of $BOD_{5, 20}$ removals, according to WEF and ASCE (1992). The degree of stabilization depends on many factors such as; volumetric and organic loading, kind of support media, temperature and ventilation.

Trickling Filter Equipment

Trickling filter tanks are generally constructed above the ground, they are either be rectangular tank with spray nozzles (Network of pipes having fixed nozzles, spray the incoming sewage into the air, sewage falls on the bed of filter media under the action of gravity) or more generally circular tank with rotatory distributor (Generally four or more distributary arms are being used in this type with a central supporting main central column with the self-rotating and mechanical rotating system, for small and large discharge the rpm should be 2 and 1/2).

Trickling filter that use plastic packing having build in round, square and other shapes with depth varying from 4 – 12 m (14 to 40 ft.), Ideal trickling filter bio film carriers or media, provide a high specific surface area, low cost, high durability and high enough porosity to avoid clogging and promote ventilation. These include; rocks (river, slag), plastics (cross flow "PVC", vertical flow "PVC", random "polypropylene"). The depth of the rock or filter media packing is around 0.9 – 2.5 m (3 to 8 ft.).

The height of the distributary system from the top surface of the filter media should be 15 - 20 cm. After dozing time of waste water is 3 - 5 minutes and after dozing the resting time is 5 - 10 minutes, the slime layer build on the media filter by the action of bacteria and under the process of degradation and the thickness can reach the depth as much as 10 mm. in the outer portion of the biological slime layer (0.1 – 0.2 mm), the organic matter is degraded by the action of aerobic microorganisms and with the increase in thickness of the slime layer, oxygen is consumed before it can penetrate the full depth, and an anaerobic environment is established near the surface of the packing. Therefore the bacteria in the slime layer enter an endogenous respiration state and lose their ability to cling to the packing surface and then liquid washes this layer and new layer of slime start to grow. The breakup or detachment of the biomass (biological solids) from the slime layer is known as sloughing and this sloughed material is separated from the treated sewage in the secondary settling tank with the help of under drainage system provided below the tank.

By properly designed under drainage system an adequate aeration can be achieved. They consist of v-shaped or semi cylindrical shaped channels made of concrete with the slop of about 1 in 100 to 1 in 50 for the collection of treated sewage from trickling filter, the system should be self-cleaning and easily handy during onslaught and should be open to the atmosphere with the vertical pipes at some places for the purpose of introducing the compressed air into the system at a velocity of 60 cm/sec for aeration (Sengupta 2017).

Traditionally a trickling filter is installed with sedimentation tank; primary sedimentation tank is including removing large settle able solid from the incoming wastewater. This the load onto the biological treatment after that resulting in less excess biomass being produces. Wastewater treatment using trickling filter result in a net production of total suspended solids, require the separation of liquid-solids with the help of secondary sedimentation tank (clarifier).

Low or standard rate TF – They are also known as standard rate or conventional rate filters. The filter may be circular or rectangular in shape, filter media of rock are used in this type of TF. A well operated low rate TF can provide good BOD removal and highly nitrified effluent. Odours are the common problem, especially if the waste water is stale or septic, or if the weather is warm. Filters should not be located where the odour would create the nuisance. Filter flies (psychoda) may breed in the filer unless effective control measures are used.

Table 8. Filter classifications

Physical Features	Low or Standard Rate (Rock)	Moderate Rate (Rock)	High Rate (Rock/ Plastic)	Super Rate (Rock/Plastic)
HLR (hydraulic loading rate) m^3/m^2*d	1 – 4	4 – 10	10 – 40 10 – 75	40 – 200
OLR (organic loading rate) kg BOD/m^3*d	0.07 – 0.22	0.24 – 0.48	0.4 – 2.4 0.6 – 3.2	>1.5
Sloughing	Intermittent	Intermittent	Continuous	Continuous
Depth, m	1.8 – 2.4	1.8 – 2.4	1.8 – 2.4 3.0 – 12.2	0.9 – 6
BOD removal efficiency	80 – 90	50 – 80	50 – 90 60 – 90	40 – 70
Type of effluent	Well nitrified	Some nitrification	No nitrification	No nitrification

High and super rate TF – plastic media of 25 to 100 mm size are being used in this the high rate type. This is mainly constructed for the treatment of wastewater having very high BOD (more than 1.6 kg/m^3*d) and higher hydraulic rate (up to 190 m^3/m^2*d). As plastic media have high specific surface area, high void space, and lower mass, the trickling filter can handle much higher hydraulic and organic loading rate, thus requiring lesser space and lesser filter media.

Standard Operation Procedure

BOD Removal

Tank with coarser permeable filter media on which the attach growth of bacteria (heterotrophic bacteria) is form over which the sewage is allow to sprinkle or trickle down, with the help of rotating distributers in the circular tank or perforated parallel pipes in the rectangular tank. Wastewater flows over plastic or rock filter media upon which an active bacterial mass (films of microorganisms are formed on filtered media) grows, using the wastewater as substrate (food) for growth and respiration, as the biological film grows or thickens the microorganism at the lower portion of the film is lose their ability to cling to media filter and some portion of the film is falls over the media filter (this process is known as sloughing). From this activity the organic carbon is transformed from being human waste into new bacterial mass. It is an aerobic waste water treatment process. The percolating sewage is collected at the bottom of the tank through a well-designed under – drainage system that collect the filtrate and solids below the tank. And piped and collected in secondary sedimentation (secondary clarifier) tank where all the solids are separated from the treated effluent, part of it diluted with the influent from the primary settling tank (primary clarifier) for process optimization. Wastewater compounds such as metals are not removed in the trickling filter.

Nitrification

The process in which the ammonia nitrogen first oxidized to nitrite nitrogen and then in nitrate nitrogen by autotrophic bacteria. Heterotrophic bacteria are more competitive than nitrifying bacteria, nitrification starts when the soluble BOD concentration is low (20 – 30 mg/l) enough for nitrifiers to compete with heterotrophs. It may be accomplished with lower organic rate to remove BOD and nitrification.

General Process Description

Design Consideration of TF

Factors that must be consider before designing of trickling filter primarily involves, the volume of the tank-area of the tank with assuming depth and diameter of the TF and number of filter units or if we have the required area for construction of TF or the diameter of the TF may be calculated. The central column with rotatory distributary system (arms) with its orifices, under drainage system with the desired slope of the channel for drainage of effluent.

1. Guidelines and Procedures of the designing
   a. **Rotary Distributer:** ½ RPM – large distributer or 2 RPM – for small distributers.
   b. **HLR:** Hydraulic loading rate
   c. **Conventional Filters:** 22 – 44 ML/ha/day
   d. **High Rate Trickling Filters:** 110 – 330 ML/ha/day
   e. **OLR:** Organic loading rate
   f. **Conventional Filters:** 900 – 2200 kgBOD_5/ha-m/day
   g. **High Rate Trickling Filters:** 6000 – 18000 kgBOD_5/ha-m/day
   h. Diameter of the rotatory distributors up to 60 m length available it is desirable to keep
   i. The diameter of filter tank 60 m maximum if diameter >60 m use two or more units.
2. Formula given by NRC (National research council) USA for efficiency calculation of trickling filter (without recycling)

$$\eta\% = \frac{100}{1 + 0.0044\sqrt{u}} \tag{49}$$

where,

η Is efficiency of filter and its secondary clarifier in terms of applied BOD (%) u is organic loading rate kg/ha-m/day (OLR)

3. Efficiency of high rate trickling filter (with recirculation)

$$F = \frac{1 + \frac{R}{I}}{\left(1 + 0.1\frac{R}{I}\right)^2} \qquad (50)$$

where,

$$\frac{R}{I} = \frac{Recirculation}{Original\,Flow}$$

$$\frac{R}{I} = Recirculation\,ratio$$

$$\frac{R}{I} = 0\left(When\,no\,recirculation\,F = 1\right)$$

4. Efficiency of single stage high rate trickling filter

$$\eta\% = \frac{100}{1 + 0.0044\sqrt{\frac{Y}{V} \times F}} \qquad (51)$$

where,

Y is organic loading rate in kg/day applied to filter (total BOD inflow)
V is filter volume in ha-m
F is recirculation factor

5. Efficiency of double stage or two stage trickling filter

$$\eta'\% = \frac{100}{1 + \frac{0.0044}{1-\eta}\sqrt{\frac{Y'}{V'}} \times F'} \quad (52)$$

where,

Y' = Total BOD influent from 1st stage (kg/day)
V' = Volume of second stage filter (ha-m)
F' = Recirculation factor for second stage
η' = Final efficiency obtained after second stage
Secondary clarifier:

a. **Detention Period:** 1.5-2 hours
b. **SOR Standard Operating Rate:** 40-70 $m^3/m^2/day$
c. **Diameter:** 7.5 m and Height: 9 m

6. Efficiency at T temperature

$$\eta_T = \eta_{20°}\left(1.035\right)^{T-20} \quad (53)$$

7. Efficiency of BOD removal

$$\eta = \frac{BOD_{in} - BOD_{out}}{BOD_{in}} \times 100$$

Where BOD_{in} is the influent BOD and BOD_{out} is the effluent BOD.

8. Oxygen required for - BOD removal and BOD removal and nitrification.

$$R_O = \left(20kg / kg\right)\left[0.80e^{-9L_B} + 1.2e^{-0.17L_B}\right]\left(PF\right)\text{BOD removal} \quad (54)$$

$$R_O = \left(40kg / kg\right)\left[0.80e^{-9L_B} + 1.2e^{-0.17L_B} + 4.6N_{OX/}BOD\right]\left(PF\right)$$

BOD and nitrogen removal

where,

R_O = Oxygen supply, kg O_2/kg BOD applied

L_B = BOD loading to filter, kg BOD/m^3*d
N_{OX}/BOD = ratio of influent nitrogen oxidized to influent BOD, mg/mg
PF = peaking factor, maximum to average load

Overview and Brief Summary of Trickling Filter Process

Trickling filter is an aerobic waste water treatment technology with an attached growth treatment process.

Advantages

1. Able to handle high organic loading rate depending on media filter.
2. Efficiently remove the ammonia from the waste water.
3. Suitable for small and medium size of communities.
4. Able to handle the heavy and shock load with the help of introduction of plastic filter media, speed control and consistent distributary system.
5. Moderately lower electrical power consumption as compare to other treatment technologies.
6. Less sludge production and well settleable sludge.
7. Skill labour and technical expertise not required for the plant and operating cost is low.

Disadvantage

1. Additional treatment of the waste water is required for achieving the standards.
2. Generated sludge must be treated for the disposal.
3. Regular attention is needed.
4. High incidence of clogging is accounted.
5. Loading rate is depends on the filter media.
6. Low flexibility and control as compare to activated sludge process.
7. Odour and fly problem and high maintenance cost of rotatory distributors.

Trickling Filter Process Calculations

Example 13

Design a circular Trickling Filter with:-

1. Trickling filter dimensions
2. Rotatory distributary system
3. Under drainage system

For 5 MLD discharge and 200 mg/L BOD load.
Sol. Trickling filter dimensions
Design of the tank on OLR
Total BOD present in the influent sewage per day

$$Q \times BOD$$

$$5MLD \times 200mg / L$$

$$BOD = \frac{1000Kg}{d}$$

1. Volume of the tank design on the basis of OLR value

Volume of the tank = BOD/OLR
Assume OLR = 1500 Kg/ha-m/d (should be in between 900-2200 Kg/ha-m/d)

$$V = \frac{100 \frac{Kg}{d}}{1500 \frac{Kg}{ha * m * d}}$$

$$V = 0.666666 ha * m$$

$$V = 6666.66 m^3$$

2. Total area required for the given BOD and OLR

Area of the filter media require = volume/depth
Assume depth of the tank 4 m

$$A = \frac{6666.66m^3}{4m}$$

$$A = 1666.66m^3$$

3. Number of filter units required = area of the media filter/area of the single unit

Assume diameter of the tank 40 m

$$\text{No. of filter units } = \frac{1666.66m^3}{\frac{\pi}{4} \times d^2}$$

$$\text{No. of filter units } = \frac{1666.66 \text{ m}^3}{\frac{\grave{A}}{4} \times 4^2}$$

$$\text{No. of filter units } = 1.3 say 2$$

4. Check for hydraulic loading rate (HLR)

Assume 25 ML/ha/d (should be in between 22-44 ML/ha/d)
Check for the area require

$$Area = \frac{Total\, area\, to\, be\, treated\, /\, day}{hydraulic\, loading\, rate\left(HLR\right)}$$

$$Area = \frac{5\, MLD}{25\frac{MLD}{ha * d}}$$

$$Area = 2000m^3$$

Check for the area chooses and calculated for the assume HLR and compare the both

$1666.66\ m^2 < 2000\ m^2$ (Not OK)

Reduce the depth of the tank for achieving the satisfy area

From the above

Total area required for the given BOD and OLR

Area of the filter media require = volume/depth

Assume depth of the tank 2 m

$$Area = \frac{6666.66m^3}{2m}$$

$$Area = 3333.33m^3$$

Number of filter units required = area of the media filter/area of the single unit

Assume diameter of the tank 40 m

$$\text{No. of filter units} = \frac{3333.33m^3}{\frac{\pi}{4} \times 2^2}$$

$$\text{No. of filter units} = 2.6 say 3$$

Check for the area chooses and calculated for the assume HLR and compare the both as above

$3333.33\ m^2 < 2000\ m^2$ (hence OK)

Therefore Dimensions of the trickling filter is:-

Three filter units are require, of depth 2 m with 0.6 m free board hence 2.6 m require depth, volume of the tank 6666.66 m^3 with surface area of 3333.33 m^2.

Rotatory distributary system

Design on the peak flow

Peak flow = Q*Peak factor

$$Peak\ flow = 5 \times 2.25$$

Peak flow = 11.25 MLD

Peak flow = 0.13 m^3/s

Therefore,

$$\text{flow in each filter unit} = \frac{0.13\frac{m^3}{s}}{2}$$

$$Q = 0.065\frac{m^3}{s}\text{ in each unit}$$

5. Area and Diameter of the central column require

Assume velocity of the flow is 2 m/s (should be in between 1.5-2 m/s)
Discharge = area*velocity

$$Area = \frac{0.065m^3 / s}{2m / s}$$

$$Area = 0.032m^2$$

$$\text{Diameter of the pipe} = \sqrt{\frac{4}{\pi}\times area}$$

$$\text{Diameter of the pipe} = \sqrt{\frac{4}{\pi}\times 0.032}$$

$$\text{Diameter of the pipe} = 0.20m$$

6. Check at the average flow

Discharge = 5 MLD/2 units

$Q = 0.029m^3 / s$

Velocity at average flow = discharge/area

$$Velocity = \frac{0.029\frac{m^3}{s}}{\frac{\pi}{4} \times 0.20^2}$$

Velocity = 0.92 m/s (not OK the minimum velocity should be 1.5 m/s)
For achieving the nominal velocity reduce the diameter of the central column
Say diameter of the central column 0.15 m

Therefore Velocity at average flow = discharge/area

$$Velocity = \frac{0.029\frac{m^3}{s}}{\frac{\pi}{4} \times 0.15^2}$$

$Velocity = 1.647\, m / s$ (Hence OK)

7. Design of distributary system with 4 distributary arms

Discharge per arm = 0.065/4
Discharge per arm = 0.016 m^3/s

a. Diameter of the filter bed used = 40 m

Therefore length of an arm = 40-2/2
Therefore each arm of 19 m length having the same flow throughout the length with reducing diameter

b. Let A_1, A_2, A_3 be the circular area of filter for providing the relative flow into the pipes for covering the whole filter area by the pipes so that in one rotation of one pipe the whole filter area can covered by the discharged sewage.

The length can be dividing into three parts for calculating the dimensions of the pipes

Say $A_1 = 7$ m+$A_2 = 7$ m+$A_3 = 5$ m = 19 m

Starting from the central column
Area of the filter portion covered by the pipe of 7 m arm:-

$$A_1 = \pi\left(r_2^2 - r_1^2\right)$$

$$A_1 = \pi\left(7.15^2 - 0.15^2\right)$$

$$A_1 = 160.45\,m^2$$

Area of the filter portion covered by another pipe of 7 m arm:-

$$A_2 = \pi\left(r_3^2 - r_2^2\right)$$

$$A_2 = \pi\left(14.15^2 - 7.15^2\right)$$

$$A_2 = 468.174\,m^2$$

Area of the filter portion covered by the pipe of 5 m arm:-

$$A_3 = \pi\left(r_4^2 - r_3^2\right)$$

$$A_3 = \pi\left(20^2 - 14.15^2\right)$$

$$A_3 = 627.30\,m^2$$

Check the total area covered by the arm without central column

$$A_1 + A_2 + A_3 = 1255.924\,m^2$$

$$A = \pi\left(20^2 - 0.15^2\right)$$

$$A = 1255.92\,m^2$$

Percentage of each area:-

$$A_1 = \frac{A_1}{A} \times 100$$

$$A_1 = \frac{160.45\,m^2}{1256\,m^2} \times 100 = 12.77\%$$

$$A_2 = \frac{A_2}{A} \times 100$$

$$A_2 = \frac{146.17\,m^2}{1256\,m^2} \times 100 = 37.27\%$$

$$A_3 = \frac{A_3}{A} \times 100$$

$$A_3 = \frac{627.30\,m^2}{1256\,m^2} \times 100 = 49.94\%$$

Total percentage = 100%
Assuming the velocity through each arm 1.2 m/s

1. Design of the first section of the pipe

$$Area = \frac{Discharge}{velocity} \times Discharge\,amount$$

$$Area\,required = \frac{0.016}{1.2} = 0.0133\,m^2$$

Therefore

$$diameter\,required = \sqrt{area \times \frac{4}{\pi}} = 0.13m\,or\,130mm$$

2. Design of the second section of the pipe

$$Area\,required = \frac{0.016}{1.2} \times \left(\frac{100 - 12.77}{100}\right) = 0.0116\,m^2$$

Therefore

$$diameter\,required = \sqrt{area \times \frac{4}{\pi}} = 0.1215m\,or\,121mm$$

3. Design of the third section of the pipe

$$Area\,required = \frac{0.016}{1.2} \times \left(\frac{100 - 37.27}{100}\right) = 0.008364\,say\,0.010\,m^2$$

Therefore

$$diameter\,required = \sqrt{area \times \frac{4}{\pi}} = 0.1128\,m\,or\,112\,mm$$

Different diameter of pipes can be used for construct the distributary pipe system.

Design of orifices of the distributary system

4. No. of orifices as per discharge

Total discharge through each arm = 0.016 m^3/s
Assume 10 mm diameter of orifices with Cd = 0.65

5. Assume water head = 1.5 m

$$Q = Cd \times A \times \sqrt{2gh}$$

$$Q = 0.65 \times \left(\frac{\pi}{4} \times 0.016^2\right) \times \sqrt{2 \times 9.8 \times 1.5}$$

$$Q = 2.768 \times 10^{-4}\, m^3 / s$$

6. Total number of orifices through each arm

$$\text{Total number of orifices through each arm} = \frac{Total\, discharge\, through\, each\, arm}{discharge\, through\, each\, orifice}$$

$$\text{Total number of orifices through each arm} = \frac{0.016\, m^3 / s}{2.768 \times 10^{-4}}$$

$$\text{Total number of orifices through each arm} = 58$$

7. Number of orifice in each section

$$\text{First section} = \frac{12.77}{100} \times 58 = 7.4\, say\, 7$$

$$\text{Second section} = \frac{37.27}{100} \times 58 = 21.61\, say\, 22$$

$$\text{Third section} = \frac{49.94}{100} \times 58 = 28.96\, say\, 29$$

$$7 + 22 + 29 = 58$$

8. Spacing between the orifices

$$\text{First section } = \frac{7}{7} = 1mc / c$$

$$\text{Second section } = \frac{7}{22} = 0.31mc / c$$

$$\text{Third section } = \frac{9}{29} = 0.31mc / c$$

9. Design of under drainage system

Total discharge through each filter unit at peak flow = 0.065 m^3/s

The block or under drainage system, is constructed as rectangular block channel with radial lateral effluent discharging system. The size and the slop should be following the criteria of velocity 1 m/s or minimum 0.9 m/s.

10. Area of the channel

$$\frac{Discharge}{Velocity} = \frac{0.065 m^3 / s}{1m / s} = 0.065 m^2 / s$$

$$Depth = \frac{Area}{width}$$

$$Depth = \frac{0.065m^2}{assume\, width\, of\, 0.225\, m\left(add\, free\, board\, 22\, cm\right)}$$

$$Depth = 0.288\, m\, say\, 0.3m\left(add\, free\, board\, 30cm\right)$$

11. Slop of the under drainage system for effluent discharge

$$Q = \frac{1}{N} \times A \times R^{\frac{2}{3}} \times S^{\frac{1}{2}}$$

Where,

Q = discharge
N = Manning's coefficient (0.018)
A = 0.225 m*0.3 m = 0.067 m^2
R = A/P (area of the channel/wetted parameter of the channel)
R = 0.067 m^2/(0.225+0.3+0.3)
R = 0.082

$$Q = \frac{1}{N} \times A \times R^{\frac{2}{3}} \times S^{\frac{1}{2}}$$

$$S = \sqrt{\frac{Q \times N}{A \times R^{\frac{2}{3}}}}$$

Therefore S = 1/117
Slope is 1 in 100 m

Rotating Biological Contactors

Rotating biological contactors (RBC), also called rotating biological filters, are fixed-bed reactors consisting of stacks of rotating disks mounted on a horizontal shaft. They are partially submerged and rotated as wastewater flows through. They are used in conventional wastewater treatment plants as secondary treatment after primary sedimentation of domestic grey- or black water, or any other biodegradable effluent. The microbial community is alternately exposed to the atmosphere and the wastewater, allowing both aeration and assimilation of dissolved organic pollutants and nutrients for their degradation.

Advantages

1. High contact time and high effluent quality (both BOD and nutrients)

Box 2.

S No.	Parameter	Value	Units
1	Volume of the tank	6666.66	m^3
2	Area of the tank	3333.33	m^2
3	Depth of the tank plus free board	2 + 0.6 = 2.6	m
4	No. of filter units	3	-
5	Diameter of the central column	0.15	m
6	Diameter of the filter bed	40	m
7	Length of the distributer pipe	19	m
8	Diameter of 1,2,3 section of pipe	130, 121, 112	m
9	Number of orifices in each section	7, 22, 29	-
10	Spacing between the orifices in each section	1, 0.31, 0.31	m c/c
11	Width and depth of the under drainage system	0.225, 0.3	m
12	Slop of the under drainage system	1 is to 100	m

2. High process stability, resistant to shock hydraulic or organic loading
3. Short contact periods are required because of the large active surface
4. Low space requirement
5. Well drainable excess sludge collected in clarifier
6. Process is relatively silent compared to dosing pumps for aeration
7. No risk of channelling
8. Low sludge production

Figure 17. Schematic design of rotating biological contactors

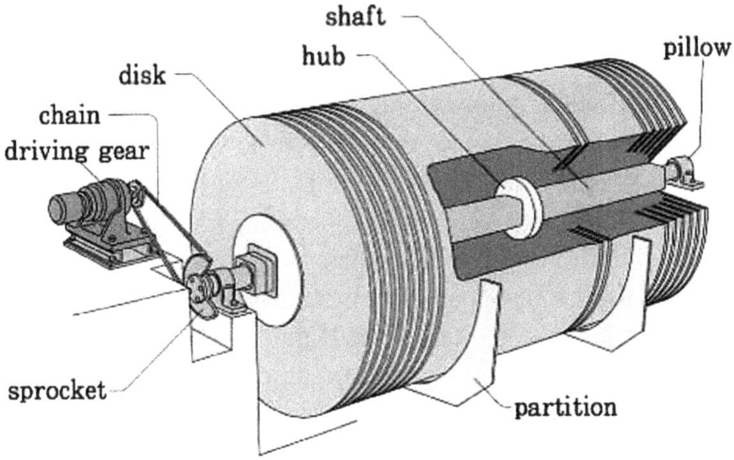

Disadvantages

1. Continuous electricity supply required (but uses less energy than trickling filters or activated sludge processes for comparable degradation rates)
2. Contact media not available at local market
3. High investment as well as operation and maintenance costs
4. Must be protected against sunlight, wind and rain (especially against freezing in cold climates)
5. Odour problems may occur
6. Requires permanent skilled technical labour for operation and maintenance

Rotating biological contactors (RBC) are a conventional aerobic biological wastewater treatment unit. Conventional biological treatment means activated sludge systems and fixed film systems such as trickling filters, or RBC. The advantage of all these systems is that they are compact (i.e. in densely populated urban settings) and that they efficiently reduce organic matter. However, they are high-tech and generally require skilled staff for construction as well as for operation.

RBC can treat domestic black- or grey water and any other low or high-strength biodegradable wastewater (e.g. industrial wastewater from food processors or paper mills). They have been found to be particularly effective for decentralised applications (on the level of a small to medium community or industry/institution), where electricity and skilled staff are available.

Treatment Process and Basic Design Principles

A series of circular lightweight rotating discs are mounted on a shaft through which wastewater flows. The partially submerged discs rotate through the wastewater slowly. The disks are most commonly made of high-density plastic sheets (e.g. Polyethylene, polystyrene or polyvinylchloride) and are usually ridged, corrugated, or lattice-like to increase the specific surface area. The surface of the disks provides an attachment site for bacteria and as the discs rotate, a film of biomass grows on their surfaces. This biofilm is alternately exposed to either the air or the wastewater as it rotates. The oxygen necessary for the growth of these microorganisms is obtained by adsorption from the air as the biofilm on the disk is rotated out of the liquid. As the biofilm passes through the liquid phase, nutrients and organic pollutants are taken up. All oxygen, nutrients and organic pollutants are necessary for the growth of the microorganism and the conversion of the organic matter to CO_2. Nitrogen is removed by nitrification and subsequent de-nitrification transforming it to gaseous N_2, which is released to the air. The process is optimised by adjusting the speed of rotation and the depth of submergence. In some designs, air is added to the bottom of the tank to provide additional oxygen in case of high-strength influents.

A series of circular lightweight rotating discs are mounted on a shaft through which wastewater flows. The partially submerged discs rotate through the wastewater slowly. The disks are most commonly made of high-density plastic sheets (e.g. Polyethylene, polystyrene or polyvinylchloride) and are usually ridged, corrugated, or lattice-like to increase the specific surface area. The surface of the disks provides an attachment site for bacteria and as the discs rotate, a film of biomass grows on their surfaces. This biofilm is alternately exposed to either the air or the wastewater as it rotates. The oxygen necessary for the growth of these microorganisms is obtained by adsorption from the air as the biofilm on the disk is rotated out of the liquid. As the biofilm passes through the liquid phase, nutrients and organic pollutants are taken up. All oxygen, nutrients and organic pollutants are necessary for the growth of the microorganism and the conversion of the organic matter to CO_2. Nitrogen is removed by nitrification and subsequent de-nitrification transforming it to gaseous N_2, which is released to the air. The process is optimised by adjusting the speed of rotation and the depth of submergence. In some designs, air is added to the bottom of the tank to provide additional oxygen in case of high-strength influents.

The performance of RBC systems depends on the design, the temperature, and the concentration of the pollutants, the rotating velocity and the hydraulic retention time. RBCs can achieve biological oxygen demand (BOD) reductions of 80 to 90%. The removal of nitrogen (which is mostly present as ammonia) by nitrification and subsequent de-nitrification is also high, because both aerobic nitrifying bacteria and anaerobic denitrifying bacteria can simultaneously live in the attached biofilm, depending on weather they are situated on the bottom of the film, close to the disc support (and thus in anaerobic or anoxic conditions) or at the top of the film exposed to the air.

Some other microorganisms which can transform ammonia (NH_3) in one single step to gaseous N_2 under anaerobic conditions have also been discovered in biofilms growing on RBC. These bacteria were called annamox and resulted in the development of innovative aerobic ammonia removal and wastewater treatment processes. Little is known about the removal of phosphorus in RBCs, but it can be presumed that large parts of the phosphorus present is either accumulated in the biofilm or in the settled and collected sludge. RBCs can be arranged in a variety of ways depending on specific effluent characteristics and the secondary clarifier design.

Excess biomass sloughs off the discs by the shearing forces exerted as the discs rotate, combined with the force of gravity. The rotation movement helps to keep sloughed solids in suspension so they can be carried to a clarifier (gravity settler) for secondary settling. The collected sludge in the clarifier requires further treatment for stabilisation, such as anaerobic digestion, composting, constructed wetlands, ponds or drying. Very often in small installations, accumulated sludge is also directed back to the septic tank for storage and partial digestion.

Effluents from RBC do not contain high levels of nutrients and are therefore not particularly interesting for agriculture, although they constitute a source of water. However, due to reduce removal of microorganisms, RBC effluents require a further treatment, such as sand filtration, constructed wetlands or another form of disinfection. RBCs are usually designed on the basis of hydraulic and organic loadings derived from pilot plants and other full-scale installation. Hydraulic retention times (HRTs) generally lye within some hours up to two days. Even though RBCs are resistant to shock loading, long-term high organic loading may cause anaerobic conditions, resulting in odour and poor treatment performance.

Recirculation is not normally practised in package fixed-film systems since it adds to the degree of complexity and is energy and maintenance intensive. However, recirculation may be desirable in certain applications where minimum wetting rates are required for optimal performance. Units may be installed at or below ground depending upon site topography and other adjacent treatment processes. Access to all moving parts and controls is required, and proper venting of the units is paramount, especially if natural ventilation is being used to supply oxygen. RBCs are often covered with a fibreglass housing to protect the disks from sunlight, wind, rain and low temperatures as performance of RCS drops considerably at air temperatures below 12°C.

Secondary Treatment- Anaerobic

Anaerobic biological treatment is well understood and used frequently as anaerobic digesters to treat complex organic solid wastes such as primary and secondary wastewater sludges. However, it has not been used much in the past to treat low strength organic wastewaters from industrial and domestic applications. Aerobic processes were preferred for treatment of these wastewater streams because they are easy to operate and can tolerate process fluctuations. In comparison, anaerobic reactors were assumed to be less stable under fluctuations, more expensive to install and require long start-up time. This belief was due to limited knowledge of the process and reactor design.

Now the technology advances have significantly reduced the historical weakness of anaerobic treatment. With the work of Young and McCarty in the year 1969, application of anaerobic process for the treatment of industrial and municipal wastewaters has gradually increased in last three decades. Today the anaerobic treatment has emerged as a practical and economical alternative to aerobic treatment due to significant advantages over aerobic treatment.

Suspended Growth Process

Suspended growth treatment systems freely suspend microorganisms in water. They use biological treatment processes in which microorganisms are maintained in suspension within the liquid. In suspended growth treatment systems, microorganisms convert the organic matter or other constituents in the wastewater into gases and cell tissue. The most common type of aerobic system is the suspended growth treatment system. Suspended growth technologies are conventional activated sludge treatment systems that use various process modes ranging from:

- Conventional
- Extended aeration
- Contact stabilization
- Sequencing batch
- Single sludge

The various process modes are available for polishing anaerobically treated effluents.

Up-Flow Anaerobic Sludge Blanket Reactor (UASBR)

It is somewhat modified version of the contact process, based on an upward movement of the liquid waste through a dense blanket of anaerobic sludge. No inert medium is provided in these systems. The biomass growth takes place on the fine sludge particles, which then develop as sludge granules of high specific gravity. The reactor can be divided in three parts (Fig 17), sludge bed, sludge blanket and three phase separator (gas-liquid-solid, GLS separator) provided at the top of the reactor. The sludge bed consists of high concentration of active anaerobic bacteria (40 – 100 g/L) and it occupies about 40 to 60% of reactor volume. Majority of organic matter degradation (> 95%) takes place in this zone. The sludge consists of biologically formed granules or thick flocculent sludge. Treatment occurs as the wastewater comes in contact with the granules and/or thick flocculent sludge.

The gases produced causes internal mixing in the reactor. Some of the gas produced within the sludge bed gets attached to the biological granules. The free gas and the particles with the attached gas rise to the top of the reactor. On the top of sludge bed and below GLS separator, thin concentration of sludge is maintained, which is called as sludge blanket. This zone occupies 15 to 25% of reactor volume. Maintaining sludge blanket zone is important to dilute and further treat the wastewater stream that has bypassed the sludge bed portion following the rising biogas. The GLS separator occupies about 20 to 30% of the reactor volume. The particles that raise to the liquid surface strike the bottom of the degassing baffles, which causes the attached gas bubbles to be released. The degassed granules typically drop back to the surface of the sludge bed. The free gas and gas released from the granules is captured in the gas collection domes located at the top of the reactor. Liquid containing some residual solids and biological granules passes into a settling chamber, where the residual solids are separated from the liquid. The separated solids fall back through the baffle system to the top of the sludge blanket.

The granular biomass from the existing UASB reactor can be used as inoculum material to start-up new UASB reactor. When such material is not available, non-granular material such as anaerobic digested sludge, waste activated sludge and cow dung manure can be used as inoculum. Granular sludge can be developed using non-granular material for inoculation. Although, there are reports of wastewaters containing high-suspended solids being successfully treated in UASB reactors without primary sedimentation, the separation of suspended solids is still suggested, especially for reactors having non-granular configuration. Pre-treatment such as sedimentation, neutralization of wastewater is normally desirable in treating waste in UASB reactor. Organic loading in the range of 1-20 kg COD /m^3.d can be applied with removal efficiency of 75 to 85% and HRT of 4 to 24 h.

Figure 18. Schematic diagram of up-flow anaerobic sludge blanket reactor

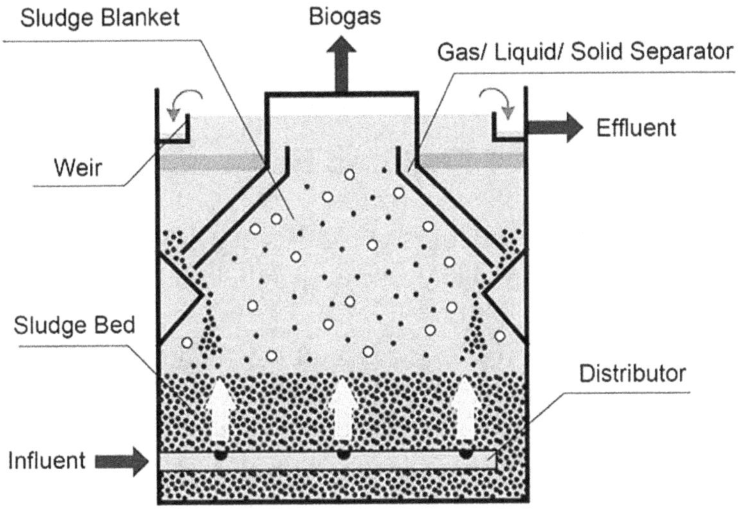

Anaerobic Contact Process

The essential feature of the anaerobic contact process is that the washout of the active anaerobic bacterial mass from the reactor is controlled by a sludge separation and recycles system. The major problem in the practical application of the contact process has always been the separation (and concentration) of the sludge from the effluent solution. For this purpose several methods have been used or were recommended for use, e.g. plain sedimentation, settling combined with chemical flocculation, with vacuum degasification, floatation and centrifugation. A basic idea underlying the contact process is that it is considered necessary to thoroughly mix the digester contents e.g., by gas recirculation, sludge recirculation, or continuous or intermittent mechanical agitation. This is generally used for concentrated wastewater treatment such as distillery wastewater.

Attached Growth Process

Attached Growth is a biological treatment process in which microorganisms responsible for conversion of organic matter or other constituents in wastewater are attached to some inert material such as: rocks, sand or specially ceramic or plastic materials. This process is also called fixed film process. Examples of Attached growth system:

- Trickling filters (biological tower)
- Rotating biological contactors (RBC)
- Packed bed reactors
- Fluidized bed biofilm reactors.

Anaerobic Packed and Fluidized Bed

Packed bed reactor (Fixed bed reactor) consists of cylinder of large diameter with multiple catalyst bed or many tubes in parallel packed with catalyst and encased in large shell.

A fluidized bed reactor (FBR) is a type of reactor device that can be used to carry out a variety of multiphase chemical reactions. In this type of reactor, a fluid (gas or liquid) is passed through a solid granular material (usually a catalyst possibly shaped as tiny spheres) at high enough velocities to suspend the solid and cause it to behave as though it were a fluid. This process, known as fluidization, imparts many important advantages to the FBR. As a result, the fluidized bed reactor is now used in many industrial applications.

The solid substrate (the catalytic material upon which chemical species react) material in the fluidized bed reactor is typically supported by a porous plate, known as a distributor. The fluid is then forced through the distributor up through the solid material. At lower fluid velocities, the solids remain in place as the fluid passes through the voids in the material. This is known as a packed bed reactor. As the fluid velocity is increased, the reactor will reach a stage where the force of the fluid on the solids is enough to balance the weight of the solid material. This stage is known as incipient fluidization and occurs at this minimum fluidization velocity. Once this minimum velocity is surpassed, the contents of the reactor bed begin to expand and swirl around much like an agitated tank or boiling pot of water. The reactor is now a fluidized bed. Depending on the operating conditions and properties of solid phase various flow regimes can be observed in this reactor.

Advantage:

1. **Uniform Particle Mixing:** Due to the intrinsic fluid-like behaviour of the solid material, fluidized beds do not experience poor mixing as in packed beds. This complete mixing allows for a uniform product that can often be hard to achieve in other reactor designs. The elimination of radial and axial concentration gradients also allows for better fluid-solid contact, which is essential for reaction efficiency and quality.

2. **Uniform Temperature Gradients:** Many chemical reactions require the addition or removal of heat. Local hot or cold spots within the reaction bed, often a problem in packed beds, are avoided in a fluidized situation such as an FBR. In other reactor types, these local temperature differences, especially hotspots, can result in product degradation. Thus FBRs are well suited to exothermic reactions. Researchers have also learned that the bed-to-surface heat transfer coefficients for FBRs are high.
3. **Ability to Operate Reactor in Continuous State:** The fluidized bed nature of these reactors allows for the ability to continuously withdraw product and introduce new reactants into the reaction vessel. Operating at a continuous process state allows manufacturers to produce their various products more efficiently due to the removal of start-up conditions in batch processes.

Disadvantage:

1. **Increased Reactor Vessel Size:** Because of the expansion of the bed materials in the reactor, a larger vessel is often required than that for a packed bed reactor. This larger vessel means that more must be spent on initial capital costs.
2. **Pumping Requirements and Pressure Drop:** The requirement for the fluid to suspend the solid material necessitates that a higher fluid velocity is attained in the reactor. In order to achieve this, more pumping power and thus higher energy costs are needed. In addition, the pressure drop associated with deep beds also requires additional pumping power.
3. **Particle Entrainment:** The high gas velocities present in this style of reactor often result in fine particles becoming entrained in the fluid. These captured particles are then carried out of the reactor with the fluid, where they must be separated. This can be a very difficult and expensive problem to address depending on the design and function of the reactor. This may often continue to be a problem even with other entrainment reducing technologies.
4. **Lack of Current Understanding:** Current understanding of the actual behavior of the materials in a fluidized bed is rather limited. It is very difficult to predict and calculate the complex mass and heat flows within the bed. Due to this lack of understanding, a pilot plant for new processes is required. Even with pilot plants, the scale-up can be very difficult and may not reflect what was experienced in the pilot trial.

5. **Erosion of Internal Components:** The fluid-like behavior of the fine solid particles within the bed eventually results in the wear of the reactor vessel. This can require expensive maintenance and upkeep for the reaction vessel and pipes.
6. **Pressure Loss Scenarios:** If fluidization pressure is suddenly lost, the surface area of the bed may be suddenly reduced. This can either be an inconvenience (e.g. making bed restart difficult), or may have more serious implications, such as runaway reactions (e.g. for exothermic reactions in which heat transfer is suddenly restricted).

Sedimentation Aided With Coagulation

Colloidal solutions that do not agglomerate naturally are called stable. This is due their large surface-to-volume ratio resulting from their very small size. In these small particles, molecular arrangements within crystals, loss of atoms due to abrasion of the surfaces, or other factors causes their surfaces to be charged.

The colloids contained in the water are negatively charged at $pH>pH_{iso}$ and positively at $pH<pH_{iso}$. These colloids are stable due to the repulsive forces between the negative charges. These colloids are destabilized by positively charged ions formed from the hydrolysis of coagulants. Destabilization of colloidal particles can be influenced by the double layer compression, adsorption and charge neutralization, entrapment in precipitates (sweep flocculation) and interparticle bridging.

Particles in colloidal size posses certain properties which prevents them from agglomeration, this may be due to the large surface to volume ratio of the particles and because of the accumulation of electrical charges on the particle surface. However the colloidal surfaces of the particles are negatively charged.

Design Criteria for Mixing and Flocculation

The design criteria for mixing and flocculation are as shown in Box 3.

Design steps for a mechanical flocculator:

1. At a given Q value assume the G values from 20 to 70 sec^{-1} and detention time'''t' accordingly. Divide the tank in 3 chambers with decreasing G values so that area of one chamber = 1/3 of total tank area.

Figure 19. Schematic diagram of destabilization of particles

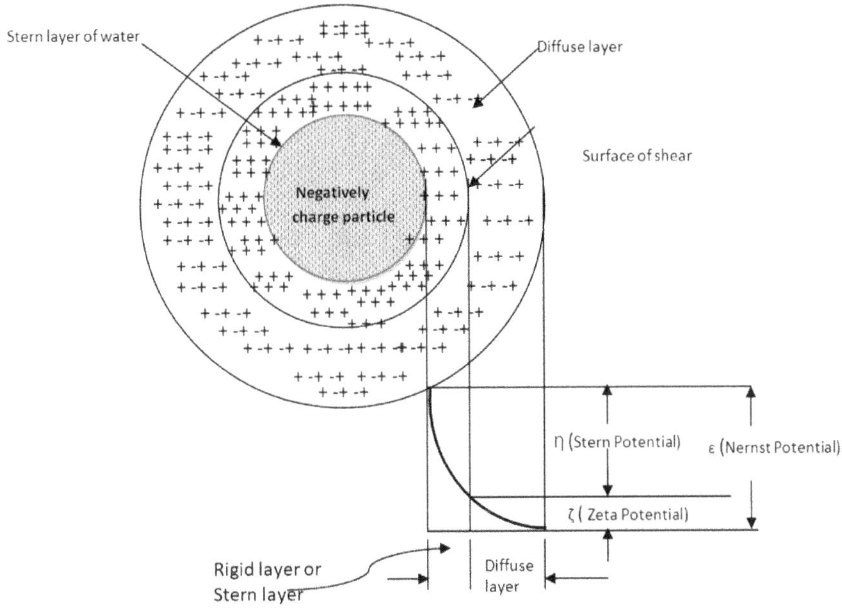

2. Designate the number of paddles per flocculator required.
3. Calculate the volume of flocculation tank using equation V = Q x t
4. Assume the depth of the tank in the range of 2-3 m and calculate the surface area of the tank by dividing the area into three parts. The tank can be divided into three chambers with decreasing G values and set the dimensions of the tank accordingly. The length can be assumed three time the width of the tank (L=3W)
5. After designating the dimensions in terms of width and length the chamber

Box 3.

Parameters	Unit	Flash Mixing	Slow Mixing (Flocculation)
Paddle Speed	rpm	100-120	2-3
Detention time	seconds	30-120	1200-3600
G t factor	dimensionless	21,000-1,20,000	10^4-10^5
Power requirement	Watts	2-5	30-70
G values	s^{-1}	700-1000	20-80

6. Set the minimum clearance of 40 cm between the outer paddle of flocculator and the floor of the tank and water surface.
7. Distance of the paddle end from the wall of the tank = 40 cm
8. Relative velocity of the paddle transferred to water (v) is taken as 75% the paddle velocity (v_p).
9. The coefficient of drag value for flat blade paddle (Cd) may be taken as 1.8.
10. Minimum spacing between the two paddles is taken as 50 cm.
11. Length of the paddle can be calculated based on the width of the chamber after deducting the minimum spacing's from both the sides of the chamber wall using equation below as:

$$\mathrm{L} = \frac{\text{length or width of the tank} - 0.5 - 0.4\text{x}2}{2}$$

12. Also assume the area of paddle = 20% of total tank area, therefore the width of the paddle can be calculated as width of the paddle = area of the paddle divided by length of the paddle.
13. Calculate the power requirement for individual chambers using the equation given as: $P_h = G^2 \times \mu \times v_h$
14. However the power delivered by the paddle can be calculated using the equation as:

$$P_p = \frac{1}{2} C_d \rho b \left(\frac{2\pi N}{60} \right)^3 \left(1 - K\right)^3 \left(r_0^{\,4} - r_0^{\,4}\right) \tag{55}$$

where, P_p = Power requirement for the paddle (W)

r_o = distance from the shat to the outer end of the paddle (m)

r_i = distance from the shaft to the inner edge of the paddle (m)

b = width of the paddle (m)

N = number of paddle rotations (RPM)

K = velocity transfer factor with a value of 0.25

Cd = coefficient of drag (value for flat blade = 1.8)

ρ = density of water (kg/m^3)

V_p= velocity of the paddle = 2πN/60 m/sec

15. However the total power requirement for the chamber can be calculated by multiplying with the total number of paddles in a chamber as $P = P_p$ x number of paddles in a chamber

Design steps for a gravity flocculator:

1. At a given Q value assume the values from 20 to 70 sec^{-1} and detention time't' accordingly. Divide the tank in 3 chambers with decreasing G values so that area of one chamber = 1/3 of total tank area. Calculate the volume of flocculation tank using equation V = Q x t
2. The area of the tank can be calculated using above volume calculated from step 1 and assuming the depth of the tank to be 2-4 m.
3. Assume length to width ratio of 3:1 and calculate the dimensions of the tank as length, width and depth.
4. The number of baffles to be provided per chamber can be calculated using equation below as:

$$n = \left[\frac{2\mu t}{\rho\left(1.44 + f\right)} \left[\frac{HLG}{Q} \right]^2 \right]^{\frac{1}{3}} \tag{56}$$

Where μ = dynamic viscosity of water (Ns/m^2)
ρ = density of water (kg/m^3)
t = detention time per compartment (sec) = total detention time/no. of compartments
f = coefficient of friction = 0.3
Q = water flow rate (m^3/sec)
L = length of the channel
G = velocity gradient (1/sec)
n= number of baffles per compartment

5. The spacing between the baffles can be calculated by dividing the length of the chamber with number of baffles (L/n)
6. The head loss per chamber can be calculated using the equation as:

$$h_L = \frac{\mu t}{\rho g} G^2 \tag{57}$$

h_L = head loss per chamber (m)
μ = dynamic viscosity of water (Ns/m^2)
ρ = density of water (kg/m^3)
t = detention time per compartment (sec) = total detention time/no. of compartments
G = velocity gradient (1/sec)
g = acceleration due to gravity (m/sec^2)
Head loss per baffle can be calculated by using equation

$$\text{head loss per baffle} = \frac{h_L}{n} \tag{58}$$

where n is the number of baffles per compartment

7. Calculate the velocity through the chamber using the head loss equation which can be equated as:

$$\frac{h_L}{n} = \frac{K}{2g}V^2 \tag{59}$$

where,
K = a constant with value of 1.5
g = acceleration due to gravity (m/sec^2)
V = velocity through the baffles (m/sec)

8. The slot width can be calculated by calculating the area of the slot using velocity and discharge through the baffle slot. Therefore the area of the slot can be calculated as:

$$Slot\,area = \frac{Q}{V}$$

where,
Q = discharge through the slot (m^3/sec)
V = velocity through the slot (m/sec)
Slot width W can be calculated as

$$W \times H = \frac{Q}{V}$$

or

$$W = \frac{Q}{V \times H} \tag{60}$$

where W = slot width (m)
H = height of the chamber
Q = discharge through the slot (m^3/sec)
V = velocity through the slot (m/sec)

The head loss and slot width in different sections can be calculated as shown in Box 4.

Chemical Precipitation and Coagulation

Chemical precipitation is the most common technique used for treatment of metal-contaminated waters. Chemical precipitation of heavy metals has long been used as the primary method of treating wastewaters in industrial applications, such as metal finishing and plating. Owing to this past success, chemical precipitation is often selected to remediate hazardous, toxic, and radioactive waste (HTRW) sites containing ground water contaminated by heavy metals or landfill leachate, or both. For the precipitation process to be effective, an efficient solids removal process must be employed. To separate the solid and liquid phases of the waste stream, coagulation, flocculation, and clarification or filtration, or both, are typically used along with precipitation. Precipitation/coagulation/ flocculation (P/C/F) systems are often used as a pre-treatment step to stop metals from interfering with subsequent treatment processes. Depending on the required treatment standards, a P/C/F system may also be used as the final stand-alone treatment.

Box 4.

Chamber No	G values (1/sec)	Number of Baffles (n)	Spacing Between the Baffles (m)	Head Loss (m)	Velocity (m/sec)	Slot Width (m)

Uses

1. **Precipitation:** Removal of dissolved substances
    a. Removal of phosphorus
    b. Removal of arsenic
    c. Removal of iron
    d. Removal of ammonium and phosphate from reject water as magnesium ammonium phosphate
2. **Coagulation-Flocculation**: Removal of colloids
    a. Removal of organics
    b. Removal of turbidity (e.g. inorganic clay)
    c. Removal of pathogenic bacteria
3. In most processes both precipitation and coagulation-flocculation happen simultaneously

Applications

1. Water treatment
2. Waste water treatment
3. Sludge and slurry dewatering
4. Water softening
5. Treating landfill leachate
6. Treating industrial wastewaters, e.g. from dairies and food processing
7. Lake restoration

Precipitation Chemicals

1. Aluminium Chloride [$AlCl_3$]
2. Poly aluminium chloride (e.g. PAX)
3. Aluminium Sulphate [alum, $Al_2(SO_4)_3$ x 14-18 H_2O] •Ferrous sulphate [$FeSO_4$x 7 H_2O] (Note, divalent iron)
4. Ferric Chloride [$FeCI_3$] – Typically sold in solution form. (e.g. PIX)
5. Ferric sulphate[$Fe_2(SO_4)_3$]•$Ca(OH)_2$

Merits and Demerits of Coagulation Process

Table 9 lists several common inorganic coagulants along with associated advantages and disadvantages.

Filtration

After screening and sedimentation the fine suspended particles (discrete, flocculated) and bacteria present in it. To reduce or to remove the remaining impurities still further to produced potable and palatable water. The water is filtered through bed of fine granular material, such as sand etc. The process of passing the water through beds of such granular materials (called filters) is known as filtration.

Filtration may help in removing odour, colour, turbidity and some pathogenic bacteria from the water. Two types of filters are generally used for treating municipal water supplies. As follows:

Table 9. Common inorganic coagulants along with associated advantages and disadvantages

Name	Advantages	Disadvantages
Aluminium Sulphate (Alum) $Al_2(SO_4)_3.18H_2O$	Easy to handle and apply; most commonly used; produces less sludge than lime; most effective between pH 6.5 and 7.5	Ads dissolved solids (salts) to water; effective over a limited pH range.
Sodium Aluminate $Na_2Al_2O_4$	Effective in hard waters; small dosages usually needed	Often used with alum; high cost; ineffective in soft waters
Poly-aluminium Chloride (PAC) $Al_{13}(OH)_{20}(SO_4)_2.Cl_{15}$	In some applications, floc formed is more dense and faster settling than alum	Not commonly used; little full scale data compared to other aluminium derivatives
Ferric Sulphate $Fe_2(SO_4)_3$	Effective between pH 4–6 and 8.8–9.2	Adds dissolved solids (salts) to water; usually need to add alkalinity
Ferric Chloride $FeCl_{3\cdot 6}H_2O$	Effective between pH 4 and 11	Adds dissolved solids (salts) to water; consumes twice as much alkalinity as alum
Ferrous Sulphate (Copperas) $FeSO_4.7H_2O$	Not as pH sensitive as lime	Adds dissolved solids (salts) to water; usually need to add alkalinity
Lime $Ca(OH)_2$	Commonly used; very effective; may not add salts to effluent	Very pH dependent; produces large quantities of sludge; overdose can result in poor effluent quality

- **Slow Sand Gravity Filters:** They can remove large percentage of impurities and bacteria from the water as compare to rapid sand filters. Slow sand gravity filters yield is very slow rate of filtration (about $\frac{1}{30}$ th of rapid gravity filters) and require large areas, and are costly. It is necessity of too much purification, and that maximum removal of bacteria has decreased. The slow sand filters are becoming obsolete days.
- **Rapid Gravity Filters:** In the modern treatment plants, rapid gravity filters are now a day almost universally adopted. Water from coagulation- sedimentation plant is directly fed into the rapid gravity filters and the resultant supplies are disinfected for complete killing of germs and colour removal.

Theory of Filtration

The filtration or purify of the water through under four different processes, as given below:

1. **Mechanical Straining**: The suspended particles in water which are bigger size than the size of voids in sand layers of the filters cannot pass through these voids and get arrested in them. The resultant water will be free from these suspended particles, most particles removed on upper layers. The arrested particles including the coagulated flocs forms a mat on the top of bed which further helps in straining out the impurities.
2. **Flocculation and Sedimentation:** Assuming that the void spaces act like tiny coagulation-sedimentation tanks; the colloidal matter arrested in voids is a gelatinous mass and therefore attract other fine particles. These finer particles thus settle down in the voids and get removed.
3. **Biological Metabolism:** Certain micro-organism and bacteria are generally present in the voids of the filters. May reside initially as coatings over sand grains or may caught during initial process of filtration. These organisms require organic impurities (such as algae, plankton etc) as their food for their survival. Organisms utilise such organic impurities and convert them into harmless compounds by the process of biological metabolism. Harmless compounds formed called schmutzdecke or dirty skin. This layer further helps in absorbing and straining out the impurities.

4. **Electrolytic Changes:** It is explained by the theory of ionisation. The filter helps in purifying the water by changing the chemical characteristics of water. Sand grains and impurities carry changes of opposite nature therefore attracts each other when comes in contact. They neutralize each other thereby changing the character of water and making it purer.

Filter Materials

Filter material like sand, sand either fine or coarse, is generally used as filter media. The layers of sand supported on gravel permit the filter water to move freely to the under drains and allows the wash water to move uniformly upward.

Sand- it's obtained from rocks like quartzite contain following properties:

1. It should be free from dirt and other impurities
2. It should be uniform in nature and size
3. It should be hard and resistant
4. It should be such as not to loose more than 5% of its weight after being placed in hydrochloric acid for 24 hours.

Size of sand: effective size

Effective size D_{10} : define as the size of the sieve in mm through which 10% of the sample of the sand by mass will pass. The size is very important because too smaller size will lead to very frequent clogging of filters and will give very low filtration rates. Too large size will permit the suspended particles and bacteria to pass through it without being removed. Uniform in size or degree of variations in sizes of particles is measured by term called uniformity coefficient.

Uniformity coefficient, i.e. ($\frac{D_{60}}{D_{10}}$) may be define as the ratio of the sieve size in mm through which 60% of the sample of the sand will pass to the effective size of sand.

Gravel: the gravel which may be used below the sand should be hard, durable, and free from impurities, proper rounded and should have density of 1600 kg/m^3.

1. **Other Materials:** Instead of using sand, sometimes, anthrafilt is used as filter media. Anthrafilt is made from anthracite, which is a type of coal stone that burns without smoke or flames. Anthrafilt- filter media

in two or multi layers have provcd vcry successful in foreign countries. This coal is cheaper, high rate of filtration and it's not available in India. Central Public Health Research Institute, Nagpur, to find out the suitability and usefulness of the bituminous coal as a substitute to anthracite.

2. **Slow Sand Filters**: The process of filtration was invented in England by James Simpson (Chelsa Water Co.) in 1829. They may still be preferred on smaller plants at warm places, where covers on filters are not required to protect the filters from freezing.

Construction of Slow Sand Filters

A typical section of slow sand filter is shown in Fig.

1. **Enclosure Tank:** Open water tight rectangular tank made of masonry or concrete. Bed slope is kept at about 1 in 100 towards, depth of tank 2.5 to 3.5m, plan area 100 to 2000 sq. m.
2. **Filter Media:** Sand layer: 90 to 110cm depth, effective size of sand (D_{10}) is 0.2 to 0.4mm, uniformity coefficient ($\frac{D_{60}}{D_{10}}$) is 1.8 to 2.5 or 3, sand layer- 15cm finer the sand purer will be water.
3. **Base Material**: Gravel 30 to 75cm thickness of depth of different sizes, depth of 3 or 4 layers- 15 to 20cm. size of gravel in the bottom layer is 40 to 65mm, size intermediate layer is 20 to 40mm, two intermediate layer used- 6 to 20mm and top most layer is 3 to 6mm.
4. **Under Drainage System:** Open jointed pipe drains or some other kind of porous drains placed 3 to 5m apart on the bottom floor and sloping toward a main covered central drain. Sometimes the main drain is placed at side of the tank.
5. **Inlet and outlet Arrangements:** Inlet chamber is constructed for admitting the effluent from the plain sedimentation tank without distributing the sand layers of the filters layer and to distribute it uniformly over the filter bed. Filter water well is also constructed on the outlet side in order to collect the filtered water coming out from the main under drain. Inlets and outlets are generally governed by automatic value.
6. **Other Appurtenances:** Vertical air pipe- depth of water above sand layer = 1-1.5m depth should not go to large variations. The loss of head caused by the resistance offered by the sand grains to the flow of water through it is usually called filter head or filtering head. It is the

difference of water levels between filter tank and the filter water wall. For a freshly unit or filter resistance is less, filter head is small 10 -15cm, loss of head: 0.7- 1.2m filter unit must be cleaned.

7. **Operation and Cleaning of Slow Sand Filters:** Back washing, Scraping & removing top sand layer: 1.5-3cm, amount of wash water: 0.2-0.6% of total water filtered. Cleaning is repeated as often as necessary until the sand depth is reduced to 40 cm, effluent after 24-36% hours should be used for the cleaning of slow sand filter.
8. **Rate of Filtration or Rate of Loading for Slow Sand Filters:** Range between: 100-200 L/hours/m^2 of filter area.
9. **Efficiency and Performance of Slow Sand Filters:** Highly efficient in removing bacteria and other suspended solids: bacteria (98- 99%) or more. These filters will also remove the odours and tastes & less efficient in removing colours. Turgidities can be removed only up to <50 mg/L. It is not suitable for turgidities >50-60 mg/L.
10. **Uses of Slow Sand Filters:** Best suited for smaller plants having: low turbidity, low colours, low bacterial contents, require huge surface area, large volume of filtering materials, very costly and uneconomical for large scale supply.

Example 13

Design six slow sand filter beds from following data:

Population to be served = 50,000 persons.

Per capita demand = 150 litres/head/day.

Rate of filtration = 180 litres/ hr/m2.

Length of each bed = Twice the breadth.

Sol. Assume maximum demand as 1.8 times the average daily demand. Also assume that one unit, out of six, will be kept as stand by.

Average daily demand = Population × per capita demand

= 50,000 × 150 litres/day

= 7.5×10^6 litres/day

Max daily demand = 1.8 × Average daily demand

= $1.8\times 7.5\times 10^6$ litres/day = 13.5×10^6 litres/day

Rate of filtration per day = (180×24) litres/m^2/day

Total surface area of filters required = $\frac{max\ daily\ demand}{rate\ of\ filtration\ per\ day}$

$$= \frac{13.5\times10^{6}\ litres\ /\ day}{180\times24\frac{litres}{m2}\ /\ day} = 3.125 \times 10^{3}\ m^{2} = 3125\ m^{2}$$

Now, six units are to be used; out of them, one is to be kept as stand by, hence only 5 units should provide the necessary area of filter required.

The area of each filter unit = $\frac{1}{5}\times$ total area required = $\frac{1}{5}\times 3125 = 625\ m^2$.

Now, if L is length and B is the breadth of each unit, then L= 2.B (given)
2B.B = 625 m^2
B^2 = 312.5 m^2
B = 17.7 m or 18 m
L = 2(18) = 36 m
Hence, use 6 filter units with one unit as stand by, each unit of size 36 m × 18 m, arranged in series with 3 units on either side.

Rapid Sand Filters or Mechanical Sand Filters

Rapid sand filters also called mechanical sand filters. Rapid sand filters are two types:

1. Utilize comparatively large size sand particles which allow greater rate of filtration as compare to slow sand filters. They are called rapid gravity filters; it's used for large municipal supplies.
2. Utilize the development of pressure over filtering water and thereby increasing the rate of filtration. They are called pressure filters; it's used for industrial plants and swimming pools.

Rapid Gravity Filters

G.W. fuller at Louisville USA coarser sand
Effective size = 0.5 mm
Yield = 30 times as SSF

Construction of Rapid Gravity Filters

1. Enclosure tank

Water tight rectangular tank: masonry concrete
Depth of tank= 2.5 – 3.5 m
Uniform distribution water: larger area
Area of filter units= 10- 80 m2 / unit
No. of units determine by (Morrell and Wallace) as below:

$$N = 1.22\sqrt{Q}$$

Where, N= no. of filter units, Q= plant capacity in MLD
There should be two filter units in any plant.
For a plant > 9 MLD

2. Filter Media

It is consists of sand layers= 60- 90 cm depth,
Effective size (D_10) = 0.35- 0.55 mm
Uniformity coefficient (D_60/D_10) = 1.3 – 1.7
Sand in layers= fine on top then coarse

3. Base Material

Gravel: Supports sand distributed the wash water also gravel= 60 -90 cm thick different sizes
5 to 6 layers each of 10- 15 cm depth
Coarsest layer= 40 mm size bottom most layer
Finest gravel= 3 mm size top most layer
Above bottom layer= 20 – 40 mm
Intermediate layers = 12- 20 mm and 6- 12 mm

4. Under Drainage System

Serves two purposes:

a. To receive and collect the filtered water
b. Allow backwashing for cleaning of filter

Backwashing

5. Wash water upward= 300- 900 L/min/m2 of filter area. Wash water should not exceed the settling velocity of smallest particle to the retained in the filter (range: 0.3- 0.9 m/min

Backwash rate = 300- 900L/min/ m2

Rate of application of wash water = 6-16 times rate of filtration (50-100L/ min/ m2).

Forms of under drainage system

1. Manifold and Lateral system
2. The wheeler bottom
3. The porous plate bottom

Manifold and Lateral System

Manifold and lateral systems are most widely used in India.

Diameter = 40 cm manifold pipe

Lateral= 10 cm diameter (placed 15 to 30 cm apart

1. In perforated pipe type of this system, lateral drains with holes in bottom side:

Diameter of holes: 6 to 13 mm

Angle of inclination: 30 degree with vertical

Spacing between holes: 7.5 to 20 cm c/c

2. **In Pipe and Strainer Type**: Strainers are placed 15 cm apart on lateral drains.

Figure 20. Plan of a manifold and lateral under drainage system for a rapid gravity filter

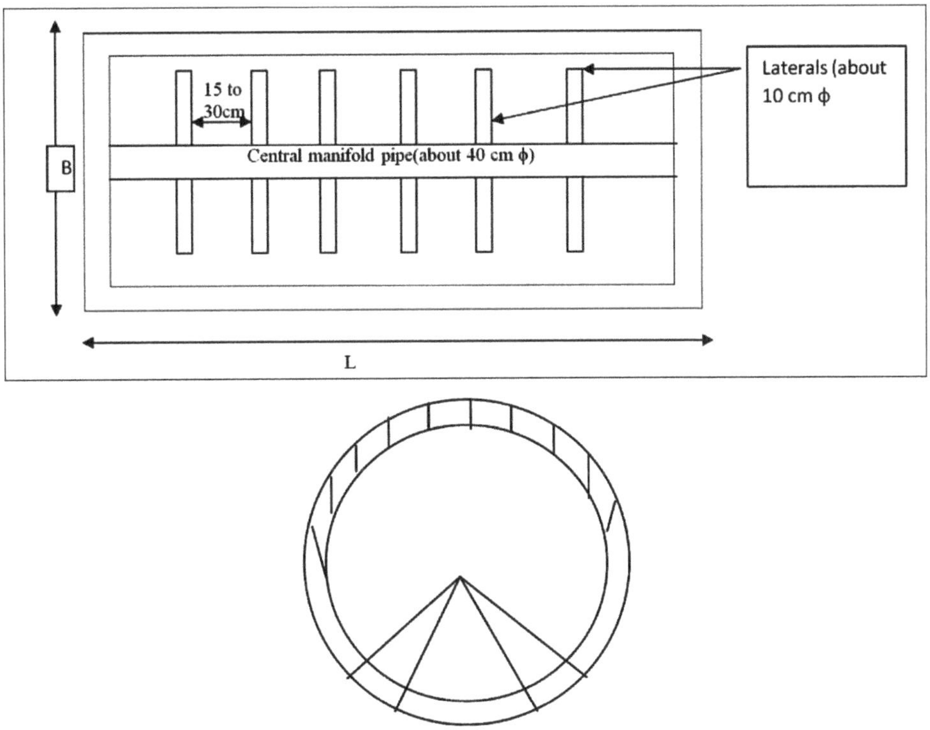

No strainer are used, the agitation of sand particles is done by water jet, backwash of high velocity required.

Rate of water application: 700 to 800 L/min/m^2 of filter area

When strainers are used, rate of water application: 250 to 300 L/min/m^2 of filter area

Design Criteria Points

1. Cross sectional area of perforation = 0.2% of total filter area
2. Cross sectional area each lateral = 2 to 4 times the cross sectional area of perforation Diameter of perforation = 13 mm to 6mm
3. Cross sectional area of the manifold = twice the cross sectional area of the lateral drains.

$$\frac{Length\ of\ each\ lateral}{diameter\ of\ the\ lateral} \not> 60$$

4. Maximum permissible velocity in the manifold required amount of wash water = 1.8 to 2.4 m/sec.

Other Appurtenances

1. Wash water trough: gutters in square, V shaped, semi-circular cast iron, and concrete, steel and rough iron.

Sand: expend 50% during backwashing overflow lip at the topes of sand rise depth above sand bed; half the depth of sand

Bottoms are adjusted: 5cm above top of sand surface

Spacing b/w the tough: 1.5m to 2m horizontal distance travelled by the wash water before entering the trough = 0.75 to 1m.

2. Air compressor: during back washing for agitation of sand grains.

Time of air supply: 4 min.
Rate of supply = 600 to 800 L/min/m^2 of filter area.
Supplied either through the lateral or through separated pipe system

3. Rate controller: Venturi rate controller, which works on the principle of venturimeter
4. Miscellaneous accessories: meters are installed for measuring discharges at entry and outlet and also at backwash.

See Box 5 for the working and cleaning of rapid gravity filters.

Box 5.

Valve Number	Name of Valve
1	Inlet valve.
2	Waste water valve to drain water from inlet chamber.
3	Waste water valve to drain water from main drain.
4	Filtered water supply valve.
5	Compressed air valve.
6	Wash water supply valve.

Back Washing

Excessive loss of head shows send becomes dirty. This forced upward movement of wash water and compressed air will agitate the sand particles and thus removing the suspended impurities from it.

Close valve 1 and 4, open valve 5 and 6, close valve 5 after required amount of air. Open valve to for the dirty water from back wash will over flow from wash water trough.

Process takes time: 3 to 5 min. close valve 2 and 6, open valve 1 and 3.

This restores the inlet supplies through valve 1 but filtered water in the beginning is not collected for few minutes and waste away through valve three to gutters. Close valve 3 and open valve 4.

The amount of water required for washing a rapid gravity filter may vary from 2 to 5% of total amount of water filter.

- **Back Washing:** 24 to 48 hrs.
- **Rate of Washing:** 15 to 90cm. rise per min.
- **Normally:** 45 cm per min.
- **Filter Rate:** 500 L/m^{2}of bed area/min.
- **Normally:** 10 to 15 minutes are used for backwash.
- **Recommissioning:** 30 minutes

Pressure of wash water supply = 40 kN/m^2

Large volume of wash water (back wash low concentration of solids) = 100 to 1000 mg/L of solids. The filter back wash may contain large fraction of organic solids filter sludge dispose after thickening.

Surface wash from top, rate of application of wash water = 200 to 600 L/ min/ m^2

Pressure = 10 to 20 L head of water (i.e. 100 to 200 kN/m^2)

Loss of Head

New filter = 15 to 30 cm

Limited value = 2.5 to 3.5 m

Negative head = 1.2 m

Depth of water during filtration increase by 15 to 20 cm

Controlling the growth of algac
Remediation by super saturated with air.
Precautions of getting water warmed.

Operation Troubles in Rapid Gravity Filter

1. Formation of mud balls.
2. Cracking of filters.

Rate of Filtration or Rate of Loading for Rapid Gravity Filters

Rate of filtration = 3000 to 6000 L/hrs/m^2 of filter area, or = 50 to 100 L/min/m^2.

Considerable saving of space as well as filter material

Efficiency and Performance of RGF

Bacterial removal = 80 to 90%
Turbidity removal = 35 to 40 NTU.
Colour removal up to 10 on cobalt scale.
Uses: Public supplies for large town and cities.

Example 14

Design a rapid sand filter unit for 4 million litre per day of supply, with all its principal components.

Sol. Water required per day = 4 M.L.

Assuming that 4% of filtered water is required for washing of the filter, every day, we have

$$\text{Total filtered water required per day} = \frac{4 \text{ M.L.}}{0.96} = 4.167 \text{ M.L/day}$$

Now, assuming that 0.5 hour is lost every day in washing the filter, we have

Filtered water required per hour = $\dfrac{4.167}{23.5}$ $\text{M.L.} / \text{hr} = 0.177\ \text{M.L} / \text{hr}$

Now, assuming the rate of filtration to be 5000 litres/hr/sq. m, we have

The area of filter required = $\dfrac{0.177 \times 10^{6}}{5000}\ \text{m}^{2} = 35.46\ \text{m}^{2}$

Now, assuming the length of filter bed (L) as 1.5 times the width of the filter bed (B), and two beds, the total area provided

$$2 \times \left(\text{L.B} \right) = 35.46$$

or

$$2 \times \left(1.5\text{B}\right)\left(\text{B}\right) = 35.46$$

or

$$\text{B}^{2} = \frac{35.46}{3} = 11.82$$

or B = 3.44 m

So, L = 1.5B = 1.5 × 3.44 = 5.16

Say 5.2 m

Or Use the length of the filter bed as = 5.2 m, and

$$\text{B} = \frac{35.46}{2 \times 5.2} = 3.4\ \text{m}.$$

Hence, adopt 2 filter units, each of dimensions 5.2 m × 3.4 m.

Design of under drainage system

Manifold and lateral system

Assume area of perforation = 0.2% of filter area = $\dfrac{0.2}{100} \times 5.2 \times 3.4 = 0.035$ m^2

Total area of laterals = 2 × total area of perforations = 2 × 0.035 = 0.070 m^2

Area of manifold = 2 × area of laterals = 2 × 0.070 = 0.14 m^2

So, Diameter of manifold (d) is given by

$$\frac{\pi}{4} \times d^2 = 0.14,$$

d = 0.42 m.

Hence, 45 cm diameter manifold pipe laid length wise along the centre of filter bottom.

Laterals emanating from the manifold at a spacing of 15 cm (c/c)

$$\text{No. of laterals} = \frac{5.2 \times 100}{15} = 35$$

Total no. of laterals = 70

$$\text{Now, length of each lateral} = \frac{\textit{Width of filter}}{2} - \frac{\textit{Dia of manifold}}{2} = \frac{3.4}{2} - \frac{0.45}{2}$$

$$= \frac{2.95}{2}$$

= 1.475m.

Now, adopt 13 mm dia of perforation in laterals,

Total area of perforations = 0.035 m^2= 350 cm^2 = x × $\frac{\pi}{4}(1.3)^2$

Where x = total no. of perforations

$$\text{So, } x = 350 \times \frac{4}{\pi} \times \frac{1}{(1.3)^2} = 264$$

So, no. of perforation in each lateral = $\frac{264}{70} = 4.0$

So, area of perforations per laterals = $4 \times \frac{\pi}{4}(1.3)^2 = 5.30\ cm^2$

Area of each lateral= $2 \times \text{area of perforations per lateral}$

$= 2 \times 5.30 = 10.60\ \text{cm2}$

Diameter of each lateral= $\sqrt{10.60 \times 4 / \pi} = 3.7\,cm$

Hence use 70 laterals each of 3.7 cm dia @ 15 cm c/c, having 4 perforations of 13 mm size with 45 cm dia manifold.

Design of Wash Water Troughs

Wash water trough= 1.5 to 2 m apart

Length of filter bed around 3 troughs= $\frac{5.2}{3} = 1.73\ \text{m aparts}$

Total wash water discharge of 0.133 m^3/s enters in these troughs.

So, discharge in troughs = $\frac{0.133}{3} = 0.044\ m^3/s$

Dimension of flat bottom trough, $Q = 1.376 b.y^{3/2}$

Where, Q= discharge m^3/s

b= width trough trough= 0.2 m
y= water depth in trough in m

$0.044 = 1.376 \times 0.2 \times y^{3/2}$

y= 0.30 m = 30 cm

Keep 5 cm freeboard the depth of trough= 30+5=35 cm

Check: $\frac{\text{length of each lateral}}{\text{dia of lateral}} = \frac{1.475\text{m}}{3.7 \text{ cm}}$ = 39.9 (< 60, ok)

Let assume rate of washing of filter 45 cm rise/min or 0.45 m/min

So, wash water discharge = $\frac{0.45 \times 5.2 \times 3.4}{60} = 0.133$

Velocity of flow in the lateral for wash water = $\frac{0.133}{70 \times \frac{\pi}{4} \times \left(\frac{3.7}{100}\right)2} = \frac{0.133 \times 10000}{70 \times 10.75}$

= 1.77 m/s

Velocity of flow in manifold = $\frac{\text{discharge}}{\text{area}} = \frac{0.133}{\pi / 4(0.45)2} = \frac{0.133}{0.158} = 0.84$ m/sec

(< 1.8- 2.4 m/s), hence O.K

Design of Filtering Sand Media for Rapid Gravity Filters

The effective size (D_{10}) for silica sand= 0.35- 0.55mm $\ngtr$1 mm

Uniformity coefficient (U) = $\frac{D_{60}}{D_{10}}$ = 1.3- 1.7

Increase of U decrease in sand uniformity grain size distribution
Coarse sand- screened
Finer sand- washed away

$$P_{use} + P_f + P_c = 100$$

Where, P_{use} - usable portion ($D_{10} - D_{60}$)

P_f - too fine
P_c - too coarse
$P_{st(10)}$– finer than P of stock sand or D_{10}
$P_{st(60)}$ –finer than P of stock sand or D_{60}

Stock of sand lies $D_{10} - D_{60}$ comprises of 50% the specified sand.

$$P_{use=2}\ P_{st(60)} - P_{st(10)}$$

Ten percent of usable sand can be below the specified D_{10} size.
% of usable stock sand D_{10} size = 0.1 P_{use}

$$P_f = P_{st(10)} - 0.1\ P_{use}$$

$P_f = P_{st(10)} - 0.2\ P_{st(60)} - P_{st(10)}$ of stock sand that is too coarse (P_c)

$$P_c = 100 - P_f - P_{use}$$

$$P_c = 100 - P_{st(10)} - 0.2\left\{P_{st(60)} - P_{st(10)}\right\} -2[P_{st(60)} - P_{st(10)}$$

$$P_c = 100 - P_{st(10)} - 1.8\ P_{st(60)} - P_{st(10)}$$

Hydraulics of Sand Gravity Filters

Initial head loss depends: Porosity (n) of filter material
Filtration velocity (v) → ½
Depth of the filter → D
Dia of the sand grain → (d)

Acceleration due to gravity → (g)
Drag coefficient→ (C_D)

Rose Equation

$$h_{L=}\frac{1.067\ v2.D}{\phi g.n4}\sum\frac{C_D.(f)}{d}$$

For uniform sand term $\sum\frac{C_D.(f)}{d}\rightarrow\frac{C_D}{d}$

$$h_{L=}\frac{1.067\ v2.D}{\phi g.n4}\sum\frac{C_D}{d}$$

Where, h_L = friction head loss through filter in m

ν = = approach velocity or filtration velocity in m/s
D= depth of filter in m
Φ = shape factor
d = dia of sand particles
g= acceleration due to gravity m/s^2
n= porosity
C_D = Newton's drag coefficient
f = mass fraction of sand particles of dia (d)
Non spherical particles d→ ϕd and $h_L \rightarrow 0.6$ m

Either loading rate too high or sand has too large proportion of fine grains.

Hydraulic Head Loss and Expansion of the Filter Bed During Backwash

To clean the interior of the filter bed it is necessary to expend it so that granules are no longer in contact with each other thus exposing all the surfaces of cleaning. To expand a porous bed the head loss or uplift must be equal to buoyant weight of the filter bed.

Head loss= h_{Le} =buoyant weight of filter bed

$$h_{Le} \cdot \gamma_w = D \cdot \gamma_{sub}$$

Where,

h_{Le} = head loss through filter bed to initiate expansion (m)
γ_w = unit weight of water (kN/m^3)
D= depth of the filter bed (m)
γ_{sub} = submerged unit weight of sand in bed at depth D (kN/m^3)

$$\gamma_{sub} = \frac{\gamma_{w(G-1)}}{1+e}$$

Where, G = specific gravity of sand grains, e = void ratio

$$\text{Porosity (n)} = \frac{e}{1+e}$$

$$1 - n = 1 - \frac{e}{1+e} = \frac{1}{1+e}$$

$$\gamma_{sub} = \gamma_w (G-1)(1-n)$$

$$h_{Le} \cdot \gamma_w = D.\gamma_w (G-1)(1-n)$$

$$\text{or } h_{Le} = D(1-n)(G-1)$$

$$\frac{h_{Le}}{D} = (1-n)(G-1)$$

When bed is expanded to depth D_e head loss remains unchanged, total buoyant weight of bed is constant.

Weight of fluidized/ expanded bed = $D_e(1-n_e)(G-1)$

So, $h_{Le} = D_e(1-n_e)(G-1)$

Where, $D_e \rightarrow$ Depth of expanded bed (m)

$n_e \rightarrow$ Porosity of expanded bed

$$D(1-n)(G-1) = D_e(1-n_e)(G-1)$$

$$\text{or } D_e = \frac{(1-n)D}{(1-n_e)}$$

These troughs should be kept at least 0.15 m above expanded bed to prevent loss of filter material.

When sand is not uniform:

$$D_e = (1-n)D.\sum\frac{f}{1-n_e}$$

Where, f= mass fraction of sand

n_e = function of terminal velocity of the particles and the backwash velocity.

$$n_e = \left(\frac{v_b}{v_s}\right)^{0.22*}$$

Where, n_e = porosity of expanded bed

v_b = Backwash velocity m/sec

v_s = Settling velocity of size d in m/s

Settling velocity (v_s) is determined by equation

$$v_s = \sqrt{\frac{\frac{4}{3}gd\left(G-1\right)}{C_D}}$$

$$R_e = \frac{v_s d\varnothing}{v}$$

Backwash rate: 15 cm/min to 90 cm/min
Normally: 45 cm/min

Pressure Filters

Closed vessels water to be treated passed under pressure. Pressure developed: 30 – 70 meter head of water or 300 – 700 kN/m^2

Construction of Pressure Filters

Diameters: 1.5 to 3.0 m
Heights: 3.5 to 8.0 m

Working Operation

Flocculation plus coagulation takes place inside these filters alum is generally used.back washing is done as per rapid gravity filters.

Rate of Filtration of Pressure Filters

2 to 5 times than RGF

Rate of filtration = 6000 to 15000 L/h/m^2 of filter area

Efficiency

Less efficient than RGF, in removing bacteria and turgidities, they may be installed for Individual house, industries, swimming pools, railway stations etc.

Advantage and Disadvantage of Pressure Filters

Advantages:

1. Compact machine, easy handling
2. Lesser space, less filter material
3. Sedimentation and coagulation tanks can be avoided.

4. Rate of filtration can be changed by pressure change.
5. Loss of head in less.

Disadvantages:

1. Capacity of plant is small.
2. Quality of effluent is poor.
3. Costlier for large scale treatment.
4. More operation and maintenance required.
5. Inspection, cleaning and replacement of sand, gravel and under drainage system is difficult.

Other Types of Filters

1. Roughening filters
2. Diatomaceous earth filters

Disinfection of Wastewater

Water disinfection means the removal, deactivation or killing of pathogenic microorganisms. Microorganisms are destroyed or deactivated, resulting in termination of growth and reproduction. When microorganisms are not removed from drinking water, drinking water usage will cause people to fall ill. Sterilization is a process related to disinfection. However, during the sterilization process all present microorganisms are killed, both harmful and harmless microorganisms.

Chlorine Disinfection

The germicidal action of chlorine is explained by the recent theory of Enzymatic hypothesis, according to which the chlorine enters the cell walls of bacteria and kill the enzymes which are essential for the metabolic processes of living organisms.

Chlorine is added to the water supply in two ways. It is most often added as a gas, $Cl_2(g)$. However, it also can be added as a salt, such as sodium hypochlorite (NaOCl) or bleach. Chlorine gas dissolves in water following Henry's Law.

$Cl_2(g)$, $Cl_2(aq)$ $K_H = 6.2 \times 10^{-2}$

Once dissolved, the following reaction occurs forming hypo-chlorous acid (HOCl):

$Cl_2(aq)+H_2O \rightarrow HOCl + H^+ + Cl^-$

Hypo-chlorous acid is a weak acid that dissociates to form hypochlorite ion (OCl^-).

$HOCl \rightarrow OCl^- + H^+$ Ka = 3.2×10^{-8}

All forms of chlorine are measured as mg/L of Cl_2 (MW = 2 x 35.45 = 70.9 g/mol)

Hypo-chlorous acid and hypochlorite ion compose what is called the free chlorine residual. These free chlorine compounds can react with many organic and inorganic compounds to form chlorinated compounds. If the products of these reactions possess oxidizing potential, they are considered the combined chlorine residual. A common compound in drinking water systems that reacts with chlorine to form combined residual is ammonia. Reactions between ammonia and chlorine form chloramines, which is mainly monochloramine (NH_2Cl), although some dichloramine ($NHCl_2$) and trichloramine (NCl_3) also can form. Many drinking water utilities use monochloramine as a disinfectant. If excess free chlorine exits once all ammonia nitrogen has been converted to monochloramine, chloramine species are oxidized through what is termed the breakpoint reactions.

The overall reactions of free chlorine and nitrogen can be represented by two simplified reactions as follows:

Mono chloramine Formation Reaction, this reaction occurs rapidly when ammonia nitrogen is combined with free chlorine up to a molar ratio of 1:1.

$HOCl + NH_3 \rightarrow NH_2Cl + HOCl$

Breakpoint Reaction: When excess free chlorine is added beyond the 1:1 initial molar ratio, mono chloramine is removed as follows:

$2NH_2Cl + HOCl \rightarrow N_2(g) + 3H^+ + 3Cl^- + H_2O$

The formation of chloramines and the breakpoint reaction create a unique relationship between chlorine dose and the amount and form of chlorine as illustrated in Figure 21.

Figure 21. Free chlorine, chloramine, and ammonia nitrogen reactions

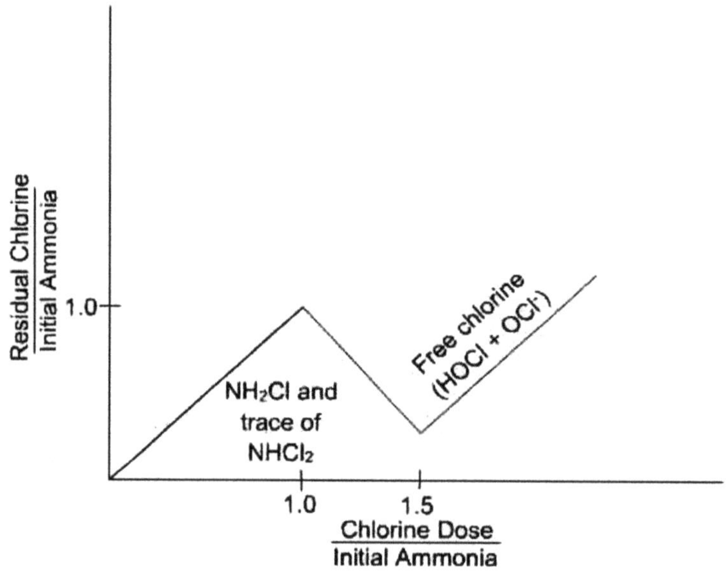

Ultraviolet Irradiation

Unlike chemical approaches to water disinfection, UV provides rapid, effective inactivation of microorganisms through a physical process. When bacteria, viruses and protozoa are exposed to the germicidal wavelengths of UV light, they are rendered incapable of reproducing and infecting. UV light has demonstrated efficacy against pathogenic organisms, including those responsible for cholera, polio, typhoid, hepatitis and other bacterial, viral and parasitic diseases. In addition, UV light (either alone or in conjunction with hydrogen peroxide) can destroy chemical contaminants such as pesticides, industrial solvents, and pharmaceuticals through a process called UV-oxidation.

Microorganisms are inactivated by UV light as a result of damage to nucleic acids. The high energy associated with short wavelength UV energy, primarily at 254 nm, is absorbed by cellular RNA and DNA. This absorption of UV energy forms new bonds between adjacent nucleotides, creating double bonds or dimers. Dimerization of adjacent molecules, particularly thymine, is the most common photochemical damage. Formation of numerous thymine dimers in the DNA of bacteria and viruses prevents replication and inability to infect.

Figure 22. Damaging of DNA of microorganism by UV radiation

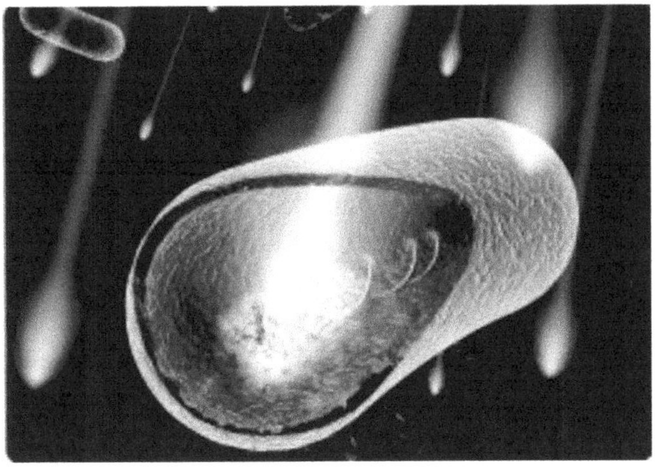

Ozonation

Because of its excellent disinfection and oxidation qualities, ozone is widely used for drinking water treatment. Ozone can be added at several points throughout the treatment system, such as during pre-oxidation, intermediate oxidation or final disinfection. Usually, it is recommended to use ozone for pre-oxidation, before a sand filter or an active carbon filter (GAC). After ozonization these filters can remove the remaining organic matter (important for final disinfection).

This combination has several benefits:

- Removal of organic and inorganic matter
- Removal of micro-pollutants, such as pesticides
- Enhancement of the flocculation/coagulation-decantation process
- Enhanced disinfection and reduction of disinfection by products
- Odor and taste elimination

Removal of organic matter and inorganic matter

All water sources contain natural organic matter (NOM). Concentrations (usually measured as dissolved organic carbon, DOC) differ from 0.2 to more than 10 mg/L. NOM creates direct problems, such as odor and taste in water, but also indirect problems such as organic disinfection by product formation, support of bacterial regrowth in the distribution system, etc. To producc pure drinking water, the removal of NOM is a prior task in modern water treatment.

Ozone, like any other oxidant, seldom achieves a complete mineralisation of NOM. Organic matter is partly oxidized and becoming more easily biodegradable. This results in a higher amount of BDOC (Biodegradable DOC). As a result, ozone improves the removal process of NOM by a subsequent filter, when it is used as a pre-oxidant. The effect of ozone in combination with a biological filter is described. The combined treatment resulted in a reduction of DOC of 40-60%. The removal is even greater when ozone is used in combination with a coagulant. This is because ozone can enhance the coagulation process. The combination coagulation–ozone–bio filtration results in a DOC reduction of 64%. When only bio filtration was apllied, the reduction rate was only 13%. The optimal concentration to remove organic matter by ozone was at an ozone dose of: O_3/DOC = 1 mg/mg.

Most inorganic matter can be eliminated by ozone quite fast. After ozonation, bio filtration is also required for inorganic matter. Namely, oxidation forms unsoluble compounds that need to be removed during the next water purification step.

One common method of disinfecting wastewater is ozonation (also known as ozone disinfection). Ozone is an un-stable gas that can destroy bacteria and viruses .It is formed when oxygen molecules (O_2) collide with oxygen atoms to produce ozone (O_3).Ozone is generated by an electrical discharge through dry air or pure oxygen and is generated onsite because it decomposes to elemental oxygen in a short amount of time. After generation, ozone is fed into a down flow contact chamber containing the wastewater to be disinfected. From the bottom of the contact chamber, ozone is diffused into fine bubbles that mix with the downward flowing wastewater. Ozone disinfection is generally used at medium- to large-sized plants after at least secondary treatment. Another common use for ozone in wastewater treatment is odor control.

Advantages:

- Ozone is more effective than chlorine in destroying viruses and bacteria.
- The wastewater needs to be in contact with ozone for just a short time (approximately 10 to 30 minutes).
- Ozone decomposes rapidly, and therefore, it leaves no harmful residual that would need to be removed from the wastewater after treatment.

- There is no regrowth of microorganisms after ozonation, unlike ultraviolet and chlorine disinfection.
- Ozone is generated onsite, and thus, there are fewer safety problems associated with shipping and handling.
- Ozonation increases the dissolved oxygen (DO) concentration of the discharged wastewater. The increase in DO can improve the oxygen content of the receiving body of water.

Disadvantages:

- Low dosages may not effectively inactivate some viruses, spores, and cysts.
- Ozonation is more complex than other disinfection technologies.
- Ozone is very reactive and corrosive, thus requiring corrosion-resistant material, such as stainless steel.
- Ozonation is not economical for poor-quality (poorly treated) wastewater.
- Ozone is extremely irritating and possibly toxic, so off-gases from the contactor must be destroyed to prevent worker exposure.
- The cost of treatment is relatively high, being both capital- and power-intensive.
- There is no measurable residual to indicate the efficacy of ozone disinfection.

Ion Exchange Process

Ion exchange is a water treatment process commonly used for water softening or demineralization, but it also is used to remove other substances from the water in processes such as de-alkalization, deionization, and disinfection. Ion exchange describes a specific chemical process in which unwanted dissolved ions are exchanged for other ions with a similar charge. Ions are atoms or molecules containing a total number of electrons that are not equal to the total number of protons. There are two different groups of ions, cations, which are positively charged, and anions, which are negatively charged. We have Michael Faraday to thank for these names, which he devised based on the cation's attraction to the cathode and the anion's attraction to the anode in a galvanic device.

Removing Ionic Contaminants

This attraction is used to remove dissolved ionic contaminants from water. The exchange process occurs between a solid (resin or a zeolite) and a liquid (water). In the process, the less desired compounds are swapped for those that are considered more desirable. These desirable ions are loaded onto the resin material. In the exchange of cations during water treatment, positively charged ions that come into contact with the ion exchange resin are exchanged with positively charged ions available on the resin surface, usually sodium. In the anion exchange process, negatively charged ions are exchanged with negatively charged ions on the resin surface, usually chloride. Various contaminants including nitrate, fluoride, sulfate, and arsenic — can all be removed by anion exchange. These resins can be used alone or in concert to remove ionic contaminants from the water. If a substance is not ionic, such as benzene, it cannot be removed via ion exchange.

Ion Exchange in Drinking Water Treatment

Recently ion exchange resins have been increasingly used to create drinking water. Specialized resins have been designed to treat various contaminants of concern, including perchlorate and uranium. There are many resins designed for these purposes, such as strong base/strong anion resin, which is used to remove nitrates and perchlorate. There are also resin beads that can be used for water softening.

Recharging Resins

Resin materials have a finite exchange capacity. Each of the individual exchange sites will become full with prolonged use. When unable to exchange ions any longer, the resin must be recharged or regenerated to restore it to its initial condition. The substances used for this can include sodium chloride, as well as hydrochloric acid, sulphuric acid, or sodium hydroxide.

The spent regenerant is the primary substance remaining from the process. It contains not only all of the ions removed, but also any extra regenerant ions and will also have a high level of total dissolved solids. It can be treated in a municipal wastewater facility, but discharges may require monitoring.

The efficacy of ion exchange for water treatment can be limited by mineral scaling, surface clogging, and other issues that contribute to resin fouling. Pre-treatment processes such as filtration or addition of chemicals can help reduce or prevent these issues.

REFERENCES

Aerosol.ees.ufl.edu. (2019). *Aerosol Transport » Section 3*. Available at: https://aerosol.ees.ufl.edu/aerosol_trans/section03_c.html

Bengtson, H. (2019). *Proportional Sutro Weir Design*. Available at: https://www.engineeringexcelspreadsheets.com/2012/11/proportional-sutro-weir-design/

CivilDigital. (2019). *Operation of Rotating Biological Contactor - Study on Sewage Treatment Plant*. Available at: https://civildigital.com/operation-rotating-biological-contactor-study-sewage-treatment-plant/

Crites, R. W., & Tchobanoglous, G. (1998). *Small and decentralized wastewater management systems*. McGraw-Hill.

Da Motta, M., Pons, M. N., Vivier, H., Amaral, A. L., Ferreira, E. C., Roche, N., & Mota, M. (2001). The study of protozoa population in wastewater treatment plants by image analysis. *Brazilian Journal of Chemical Engineering*.

Drakos, N. (1995). *Mass Transport Processes*. University of Leeds.

Environmental Protection Agency Ireland. (1995). *Wastewater Treatment Manuals. Preliminary Treatment*. Author.

GhangrekarM. M. (n.d.). Retrieved from https://scetcivil.weebly.com/uploads/5/3/9/5/5395830/m13_l18-grit_chamber-contd.pdf

Goel, R. K., Flora, J. R. V., & Chen, J. P. (2010). Flow Equalization and Neutralization. Physicochemical Treatment Processes. doi:10.1385/1-59259-820-x:021

Ho, L. T., Van Echelpoel, W., & Goethals, P. L. M. (2017). Design of waste stabilization pond systems: A review. *Water Research*, *123*, 236–248. doi:10.1016/j.watres.2017.06.071 PMID:28672208

Hoffmann, H., Platzer, C., Winker, M., & von Münch, E. (2011). *Technology review of constructed wetlands: Subsurface flow constructed wetlands for greywater and domestic wastewater treatment.* Sustainable Sanitation - Ecosan Program.

Itriwater.org.tw. (2019). *Modified Upflow Anaerobic Sludge Bed (UASB) Process--Service and R&D of innovative water technology Website.* Available at: http://www.itriwater.org.tw/Eng/technology/More?id=96

Jamal, H. (2017). *Chrystaller Central Place Theory: Assumptions, Definitions, Principles.* Retrieved from www.aboutcivil.org/preliminary-treatment-process-of-waste-water.html

Kayombo, S., Mbwette, T. S. A., Katima, J. H. Y., Ladegaard, N., & Jorgensen, S. E. (2010). *Waste stabilization ponds and constructed wetland design manual.* UNEP International Environmental Technology Center. Retrieved from http://www.unep.or.jp/Ietc/Publications/Water_Sanitation/ponds_and_wetlands/Design_Manual.pdf

Martinez, F. C., Cansino, A. T., Garcia, M. A. A., Kalashnikov, V., & Rojas, R. L. (2014). Mathematical analysis for the optimization of a design in a facultative pond: Indicator organism and organic matter. *Mathematical Problems in Engineering, 2014*, 1–12. doi:10.1155/2014/652509

Omafra.gov.on.ca. (2019). *Aeration of Liquid Manure.* Available at: http://www.omafra.gov.on.ca/english/engineer/facts/04-033.htm

Pescod, M. B. (1992). Wastewater Treatment and Use in Agriculture - FAO Irrigation and Drainage Paper 47. *International Rice Commission Newsletter, 48.* Retrieved from www.fao.org/3/T0551E/t0551e00.htm

Picswe.com. (2019). *Floating Aerators.* Available at: https://www.picswe.com/pics/floating-aerators-45.html

Primary Treatment. (n.d.). Retrieved from http://www.wefnet.org/OTWT/Chapter%203%20%20Primary%20Treatment.pdf

Quiroga, F. J. (2011). *Waste stabilization ponds for waste water treatment, anaerobic pond.* Retrieved from http://home.eng.iastate.edu/*tge/ ce421-521/Fernando%20J.%20Trevino%20Quiroga.pdf

R., & C. (2018, January 22). *Understanding Alkalinity*. Retrieved from https://www.tpomag.com/editorial/2014/05/understanding_alkalinity

Sengupta, A. (2017). Construction Features of Trickling Filters | Filtration | Sanitary Engineering." *Engineering Notes India*. Retrieved from www.engineeringenotes.com/sanitary-engineering/construction-features-of-trickling-filters-filtration-sanitary-engineering/17343

Singh, A. (n.d.). Design of primary sewage treatment plant. *Academia.edu - Share Research*. Retrieved from www.academia.edu/6078256/DESIGN_OF_PRIMARY_SEWAGE_TREATMENT_PLANT

Slideshare.net. (2019). *L 10 sedimentation*. Available at: https://www.slideshare.net/jshrikant/l-10-sedimentation

Socratic.org. (2019). *What are the four phases of the bacterial growth curve?* Available at: https://socratic.org/questions/what-are-the-four-phases-of-the-bacterial-growth-curve

Sustainabilitymatters.net.au. (2019). *Getting tired of your aging brush rotors?* Available at: https://www.sustainabilitymatters.net.au/content/wastewater/sponsored/getting-tired-of-your-aging-brush-rotors--1213900234

Technologies, T. (2019). *Introduction to UV Disinfection - TrojanUV*. Available at: https://www.trojanuv.com/uv-basics

Toprak Home Page. (2019). Available at: http://web.deu.edu.tr/atiksu/ana52/ani4053.html

United States Environmental Protection Agency. (2011). *Principles of design and operations of wastewater treatment pond systems for plant operators, engineers, and managers*. EPA 600-R-11-088. Office of Research and Development, Cincinnati. Retrieved from http://www. epa.gov/ordntrnt/ORD/NRMRL/lrpcd/projects/ponds.htm

U.S. EPA. (2000). *Wastewater Technology Fact Sheet: Screening and Grit Removal*. United States Environmental Protection Agency. (832-F-99-062)

Waste Stabilization Ponds – SSWM. (2004). *IRC International Water and Sanitation Centre*. Retrieved from https://sswm.info/sites/default/files/reference_attachments/VARON%202004%20%20Waste%20Stabilistion%20Ponds.pdf

Wastewater.com. (2019). *EDI | Fine Bubble Aeration Systems*. Available at: http://www.wastewater.com/aeration-systems/fine-bubble

Your Article Library. (2019). *Skimming Tanks: For Waste Water Treatment (explained with diagram)*. Available at: http://www.yourarticlelibrary.com/essay/skimming-tanks-for-waste-water-treatment-explained-with-diagram/29318

Chapter 4
Digestion and Disposal of Primary and Secondary Sludge

ABSTRACT

Sewage sludge is the solid, semisolid, or slurry residual material that is produced as a byproduct of wastewater treatment processes. This residue is commonly classified as primary and secondary sludge. Primary sludge is generated from chemical precipitation, sedimentation, and other primary processes, whereas secondary sludge is the activated waste biomass resulting from biological treatments. Quite often the sludges are combined together for further treatment and disposal. Sludge from biological treatment operations is sometimes referred to as wastewater biosolids. Of the constituents removed by the treatment, solids and biosolids are by far the largest in volume, and their processing, reuse, and disposal present perhaps the most challenging environmental problem and complex problem in wastewater treatment processes. Therefore, the chapter is devoted to the discussion of the sources, characteristics, quantities, disposal, digestion, and stabilization of sludge so as to present background data and information on these topics that will serve as a basis for the designing of sludge processing, treatment, and disposal facilities.

DOI: 10.4018/978-1-5225-9441-3.ch004

INTRODUCTION

The residue that accumulates in sewage treatment plants is called sludge (biosolids). Sludge is the solid material removed during the treatment of wastewaters. There are three kinds of sludge i.e sewage sludge from municipal treatment works, septage pumped from septic tanks, and industrial sludges. All three are a growing management problem throughout the world (Hope, 1986). The cost of sludge treatment constitutes to approximately half of the cost of wastewater treatment, and the quantities continue to increase as new wastewater treatment facilities are built and the existing ones are upgraded to keep up with the growing population and stricter regulations that require more treatment.

Before sludge can be disposed, it needs to be treated to a certain degree. The type of treatment needed depends on the disposal method proposed. There are principally three final disposal strategies for wastewater sludge and sludge components even though there are many "grey zones" between these are clear-cut alternatives. Sludge and sludge components may be deposited on land (in landfills or special sludge deposits), in the sea (ocean disposal) or to a certain extent in the air (mainly as a consequence of incineration) (Garg 2009).

Several studies reported that at present sludge is not seen as a waste but as a renewable source of energy, nutrients, organic matter and water. Building sludge-derived resource recovery systems will help to produce environmentally benign products, reduce dependency on nonrenewable resources, and thus facilitate the conservation of natural resources, decrease human health risk and environmental pollution, and offers the routes to the sustainable management of waste sludge (Tyagi and Lo, 2016). Resource recovery will continue to be at the center of sludge treatment and management with special emphasis on harvesting the energy value through anaerobic digestion and thermal processes, co-digestion of sludge with food and other organic waste to increase biogas production, phosphorus recovery, and beneficial use for agriculture. Agricultural use of sludge will likely be more challenging in the near future and there will be a shift towards exceptional quality sludge for land application with strict limits on heavy metals, emerging contaminants and pathogens. A shift towards wastewater treatment processes that generate less sludge either by employing pretreatment technologies or by switching from aerobic to anaerobic treatment processes is also expected. Furthermore, there is also a need for new and innovative thickening and dewatering technologies. After all, 95-99% of sludge is water and it is the water content that determines the size and design of downstream treatment processes as well as the feasibility of land application, incineration and landfilling.

SLUDGE AND ITS MOISTURE CONTENT

Primary Sludge

Primary sludge is produced through the mechanical wastewater treatment process. It occurs after the screen and the grit chamber and consists of unsolved wastewater contaminations. The sludge amassing at the bottom of the primary sedimentation basin is also called primary sludge. The composition of this sludge depends on the characteristics of the catchment area. Primary sludge consists to a high portion of organic matters, as faeces, vegetables, fruits, textiles, paper etc. The consistence is a thick fluid with a water percentage between 93% and 97%. Sludge settled in primary settling tanks comes under this category which contains 3% to 7% solids out of which approximately 60% to 80% are organic. Primary sludge solids are usually gray in color, slimy, fairly coarse, and with highly obnoxious odors. This sludge is difficult to dewater without treatment, hence digestion is necessary. This type of sludge can be digested readily by aerobic or anaerobic bacteria under favorable operating conditions.

Secondary Sludge

Secondary sludge, also known as biological sludge, is produced by biological treatment processes such as activated sludge, membrane bioreactors, trickling filters and rotating biological contactors. The sludge will contain those solids that were not readily removed by primarily clarification. Biological Sludge is more difficult to dewater than primary sludge because of the light biological flocs inherent in biological sludge (Turovskiy and Mathai, 2006). This type of sludge from secondary settling tanks has commonly a brownish, flocculent appearance and an earthy odor. It consists mainly of microorganism containing 75% to 90% organic fraction and remaining inert materials. The organic matter may be assumed to have a specific gravity of 1.01 to 1.05, depending on its source, whereas the inorganic particles have high a specific gravity of about 2.5.

Tertiary Sludge

The nature of sludge from the tertiary (advanced) treatment process depends on the unit process for example membrane processes or chemical methods, etc. Chemicals are used in wastewater treatment, espencially in industrial wastewater treatment, to precipitate and remove hard to remove substances, and in some instances, to improve suspended solids removal. In all such chemical sludge are formed (Turovskiy and Mathai, 2006). Chemical sludge from phosphorus removal is difficult to handle and treat. Tertiary sludge from biological nitrification and denitrification is similar to waste activated sludge.

Relation Between Solids Levels and Moisture Content

The relation between the level of dry solids and the moisture content in the sludge is given by:

Moisture Content (%) = 100 - Dry Solid level (%) (1)

A sludge with a level of dry solids of 3% has a water content of 97%. Therefore, in every 100kg of Sludge, 97 kg are water and 3kg are solids.The water content influence the mechanical properties of the sludge and these influence the handling processes and the final disposal of the sludge.

Waste solids production in primary and secondary processing can be estimated using the calculation below.

$$W_s = W_{sp} + W_s \quad (2)$$

where W_s = Total dry solids, kg/day

W_{sp} = Raw primary solids (kg/day) = f x SS x Q (3)
where f = fraction of suspended solids removed in primary settling
SS = suspended solids in unsettled wastewater, mg/L
Q = daily wastewater flow, ML/d

W_{ss} = secondary biological solids, kg/day = (k x BOD + 0.27SS) Q (4) where k = fraction of applied BOD that appears as excess biological growth in waste activated sludge or filter humus, assuming about 30 mg/L of BOD and suspended solids remaining in the secondary effluent

BOD = concentration in applied wastewater, mg/L

Q = daily wastewater flow, ML/d

The coefficient k is a function of process food/microorganism ratio and biodegradable (volatile) fraction of the matter in suspension. The k for secondary activated sludge processes can be estimated using Figure 1 by entering the diagram along the ordinate with a known food/microorganism ratio.

Figure 1. Hypothetical relationship between the food - to - microorganism ratio and the coefficient k in Equation 4
(S.K Garg, 2015).

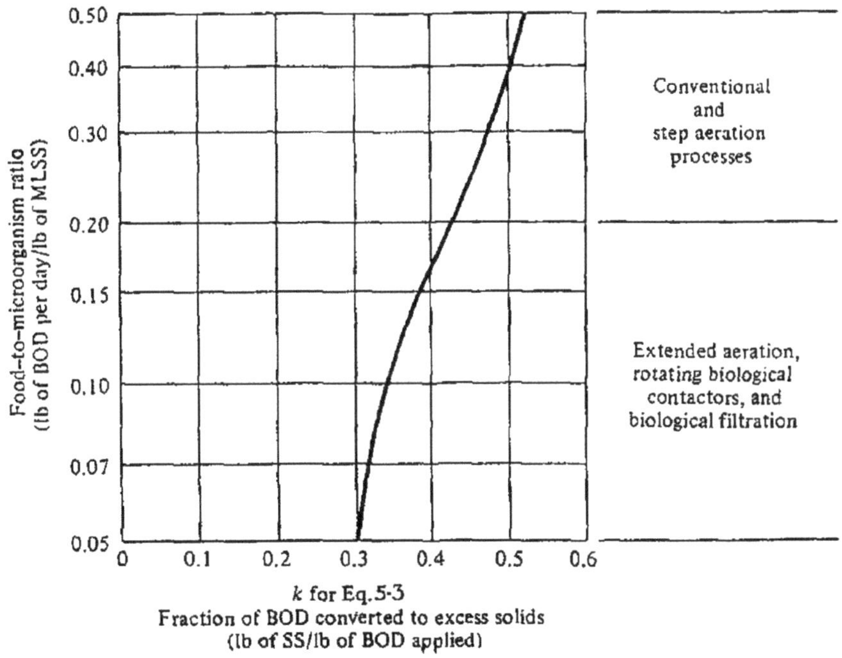

Sludge Digestion Process

Sludge digestion is a biological process in which organic solids are decomposed into stable substances. Digestion reduces the total mass of solids, destroys pathogens, and makes it easier to dewater or dry the sludge. Digested sludge is inoffensive, having the appearance and characteristics of a rich potting soil. Digestion processes may be anaerobic or aerobic. Anaerobic digestion is the main sludge stabilization process used worldwide

Main Requisites for Sludge Digestion

Normally, the presence of macro- and micronutrients is sufficient for ensuring the development of the anaerobic digestion process. The following requisites need to be observed (Table 2).

Table 1. Comparison between digested sludge and raw sludge

Raw Sludge	Digested Sludge
Unstable organic matter	Stabilized organic matter
High biodegradable fraction in organic matter	Low fraction of biodegradable organic matter
High potential for generation of odours	Low potential for generation of odours
High concentration of pathogens	Concentration of pathogens lower than in raw sludge

Table 2. Main requisites for sludge digestion

Preliminary Treatment	The sludge coming from the primary sedimentation tank contains plastics, sand, and other inert material. Such materials may cause obstruction and breakage of pipes, damage to pump rotors and to digesters mixing devices
	The accumulation of sand and other materials within the digester will end up by reducing the digester net volume and, as consequence, its efficiency.
Solids Concentration	Desirable solids concentration: 4% to 8%. Solids concentration lower than 2.5% may have negative effect on the digestion process.
	Higher solids concentrations can be used, as long as the feeding and mixing units are able to handle the solids increase.
Inhibiting Substances	Anaerobic bacteria are sensitive to several substances The main inhibiting agents are hydrocarbons, organochlorinated compounds, non-biodegradable anionic detergent, oxidizing agents and inorganic cations.
Metals	Metals like copper, zinc, mercury, cadmium, chromium, nickel and lead. excluding cadmium and mercury, the other metals are considered micronutrients if present in adequate concentrations.

(Andreoli et al, 2007)

Aerobic Digestion

The process uses organic matter, nutrients, and dissolved oxygen, and produces stable solids, carbon dioxide, and more organisms. The microorganisms which can only survive in aerobic conditions are known as aerobic organisms. This is the natural biological degradation and purification process in which bacteria that thrive in oxygen-rich environments break down and digest the waste. The end products of an aerobic process are primarily carbon dioxide and water which are the stable, oxidised forms of carbon and hydrogen. Of all the biological treatment methods, aerobic digestion is the most widespread process that is used throughout the world. Under aerobic conditions, bacteria rapidly consume organic matter and convert it into carbon dioxide. Once there is a lack of organic matter, bacteria die and are used as food by other bacteria. This stage of the process is known as endogenous respiration. Solids reduction occurs in this phase. Because the aerobic digestion occurs much faster than anaerobic digestion, the capital costs of aerobic digestion are lower. However, the operating costs are characteristically much greater for aerobic digestion because of energy costs for aeration needed to add oxygen to the process.

Advantages:

- Aerobic treatment usually yields better effluent quality that that obtained in anaerobic processes.
- The aerobic pathway also releases a substantial amount of energy.
- A portion is used by the microorganisms for synthesis and growth of new microorganisms.
- Aerobic bacteria are very efficient in breaking down waste products.

Disadvantages:

- a high power cost is associated with supplying the required oxygen
- a digested sludge is produced with poor mechanical dewatering characteristics
- the process is affected significantly by temperature, location, and type of tank material
- a useful by - product such as methane is not recovered
- More Sludge produced for disposal

Anaerobic Digestion

It is a digestion process that is characterized by the stabilization of organic matter in an oxygen-free environment. Anaerobic digestion is a multi-stage biochemical process, capable of stabilizing different types of organic matter. The process occurs in three stages:

- Enzymes break down complex organic compounds, such as cellulose, proteins and lipids, into soluble compounds, such as fatty acids, alcohol, carbon dioxide and ammonia.
- Microorganisms convert the first-stage products into acetic and propionic acid, hydrogen, carbon dioxide, besides other low-molecular weight organic acids.
- Two groups of methane-forming organisms take action: one group produces methane from carbon dioxide and hydrogen, while a second group converts the acetates into methane and bicarbonates.

Stages in Sludge Digestion Process

The overall conversion process of complex organic matter into methane and carbon dioxide can be divided into four steps as shown in Figure 2, namely hydrolysis, acidogenesis, acetogenesis and methanogenesis. The products produced by one group of bacteria serve as the substrates for the next step.

- **Hydrolysis:** Transformation of higher molecular-mass compounds into compounds suitable for use as a energy and carbon source.
- **Acidogenesis:** Bacterial conversion of the hydrolyzed compounds into identifiable lower molecular-mass intermediate compounds.
- **Acetogenesis:** Lower chain volatile fatty acids produced during acidogenesis are utilized by a acetogens to produce hydrogen and acetate.
- **Methanogenesis:** Bacterial conversion of the hydrogen and acetate into final end products, methane and carbon dioxide

Three distinct stages have been found to occur in the biological action involved in the natural process of sludge digestion. These stages are: (i) Acid fermentation (ii) Acid regression (iii) Alkaline fermentation

Figure 2. Sludge Digestion Process

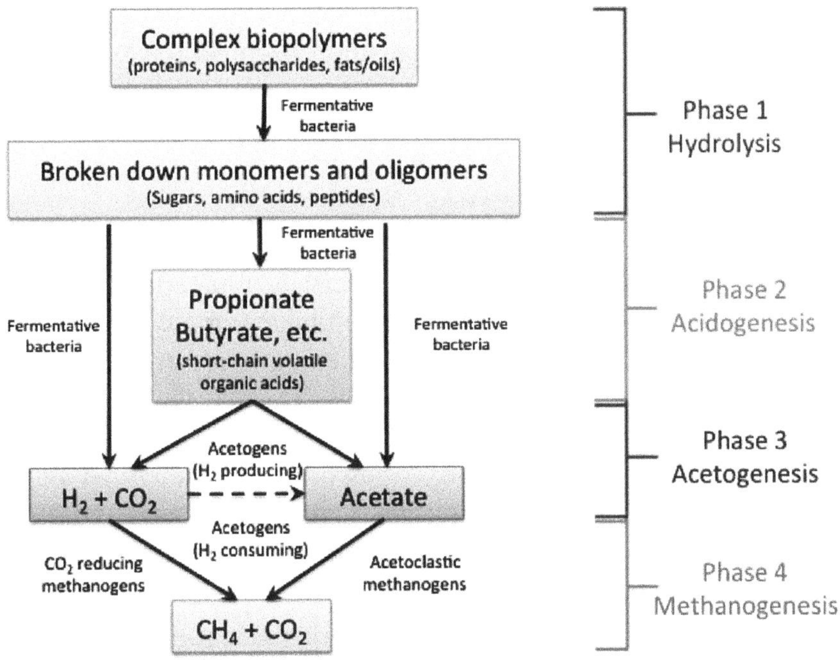

Acid Fermentation Stage or Acid Production Stage

In this first stage of sludge digestion, the fresh sewage-sludge begins to be acted upon by anaerobic and facultative bacteria, called acid formers. These organisms solubilize the organic solids through hydrolysis. The soluble products are then fermented to volatile acids and organic alcohols of low molecular weight like propionic acid, acetic acid, etc. Gases like methane, carbon dioxide and hydrogen sulphide are also evolved. Intensive acid production makes the sludge highly acidic, and lowers the pH value to less than 6. Highly putrefactive odors are evolved during this stage, which continues for about 15 days or so (at about 21°c). BOD of the sludge increases to some extent, during this stage.

Acid-Regression Stage

In this intermediate stage, the volatile organic acids and nitrogenous compounds of the first stage are attacked by the bacteria, so as to form acid carbonates and ammonia compounds. Small amounts of hydrogen sulphide and carbon dioxide gases are also given off. The decomposed sludge has a very offensive odor, and its pH value rises between 6 and 7 (Sharma 2007). The decomposed sludge, also, entraps the gases of decomposition, becomes foamy, and rises to the surface to form scum. This stage continues for a period of about 3 months or so (at about 21 °C). BOD of the sludge remains high even during this stage.

Alkaline Fermentation Stage

In this final stage of sludge digestion, more resistant materials like proteins and organic acids are attacked and broken up by anaerobic bacteria, called methane formers, into simple substances like ammonia, organic acids and gases. During this stage, the liquid separates out front the solids, and the digested sludge is formed. This sludge is granular and stable, and does not give offensive odors. (It has a musty earthy odor). This digested sludge is collected at the bottom of the digestion tank, and is also called ripened sludge. Digested sludge is alkaline in nature. The pH value during this stage rises to a little above 7 (about 7.5 or so) in the alkaline range. Large volumes of methane gas (having a considerable fuel value) along with small amount of carbon dioxide and nitrogen, are evolved during this stage. This stage extends for a period of about one month or so (at about 21°c). The BOD of the sludge also rapidly falls down during this stage.

It is, thus, seen that several months (about 4 months or so) are required for the complete process of digestion to take place under natural uncontrolled conditions at about 21°c. This period of digestion is, however, very much dependent upon the temperature of digestion, and other factors. If these factors are controlled, quicker and effective digestion can be brought about, as discussed below.

Factors Affecting Sludge Digestion and Their Control

The important factors which affect the process of sludge digestion, and are, therefore, controlled in a digestion tank, are: (i) Temperature (ii) pH value (iii) Seeding with digested sludge (iv) Mixing and stirring of the raw sludge with digested sludge.

Besides these important factors, certain other minor conditions like quality of water supply; presence of copper, fluorides, and radioactive substances, etc., may also affect the rate of digestion, but not to any appreciable extent.

These important factors which are largely responsible for controlling the rate and effectiveness of sludge digestion are discussed below:

Temperature

The process of digestion is greatly influenced by temperature; rate of digestion being more at higher temperatures and vice-versa. The effect of temperature on digestion period is shown in Figure 3.

In this figure, two distinct temperature zones are indicated; i.e.

Zone of Thermophilic Digestion

In this zone of high temperature, digestion is brought about by heat loving thermophilic organisms. The temperature in this zone ranges between 40 to 60°C. The optimum temperature in this zone is about 54°C, and at this temperature, the digestion period can be brought down to about 10 - 15 days only. However, thermophilic range temperatures are generally not employed for digesting sewage sludge, owning to odors and other operational difficulties.

Figure 3. Showing effect of temperature on sludge digestion period
(S.K Garg, 2015)

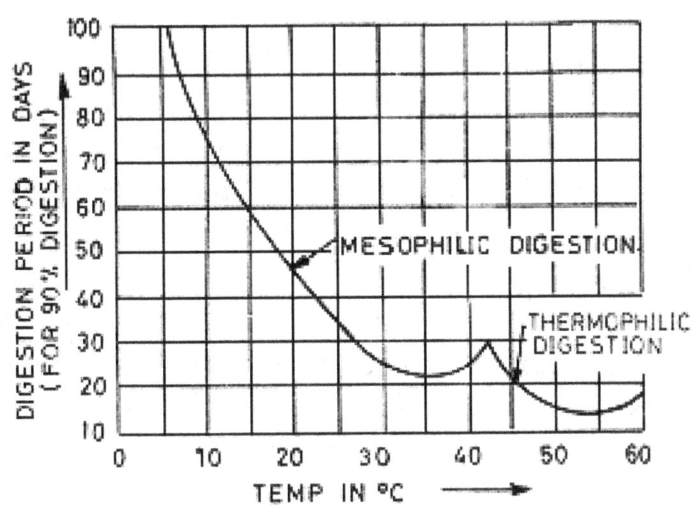

Zone of Mesophilic Digestion

In this zone of moderate temperature, digestion is brought about by common mesophilic organisms. The temperature in this zone ranges between 25 to 40°C. The optimum mesophilic temperature is about 29°C; and at this temperature, the digestion period can be brought down to about 30 days. Hence, it can be concluded that the sludge can be quickly digested, if the temperature in the digestion tank is kept high. The best results are obtained at about 29°C (i.e. the optimum mesophilic temperature) when about 90% of digestion takes place in about 30 days. But it may, however, be pointed out that it is difficult to control temperature in practice, as it mainly depends upon the prevailing local climatic conditions.

pH Value

It was pointed out earlier that during the digestion process, a lot of volatile organic acids are formed, as an intermediate step, in the breakdown of organic material. These volatile acids are then converted into methane gas by a specialized group of strictly anaerobic and slow growing bacteria, called methane formers. If the methane formers are not operating properly, an accumulation of volatile acids may occur, causing the pH to drop to a value as low as 5.0, which will (suppress further bacterial action. Hence, during digestion, care must be taken to keep the acidity well under control, so that the pH during the digester start-up does not go below 6.5 or so, and thus to see that alkaline conditions (with optimum pH about 7.2 to 7.4) may prevail ultimately, in the final stage of digestion.

The acidity increases (a) with the overdosing of raw sludge; (b) with the over withdrawal of digested sludge; and (c) with the sudden admission of industrial wastes. The remedy in such cases is to add hydrated lime in doses of 2.3 to 4.5kg/1000 persons to the raw sludge. The weight of raw sludge to be added daily, for the maintenance of optimum value of pH, should also be limited to 3 to 5 per cent of the weight of the digested sludge removed.

Seeding With the Digested Sludge

In wastewater treatment, seed, seed culture, or seed sludge refer to a mass of sludge that contains populations of microorganisms. When a sludge digestion tank is first put in operation, it is highly beneficial to seed it with the digested sludge from another tank because the successful operation of an anaerobic digestor requires the activity of an abundant and diverse population of methane forming bacteria. The raw sludge is immediately brought into close contact with the anaerobes and the active sludge acts as buffer, so preventing acid conditions at any point within the tank. To seed an anaerobic digestor with an adequate population of facultative anaerobes and anaerobes including methane forming bacteria, a ration of 1:10 of secondary sludge to primary sludge may be used. Although the amount of secondary sludge is much less compared to primary sludge, the secondary sludge is highly concentrated with facultative anaerobes. The primary sludge provides not only some facultative anaerobes but also many anaerobes including methane forming bacteria and many organic particulates (Gerardi, 2003). Without seeding, it may take a few months to get a tank operating properly. Proper seeding will help attain quick balance conditions of reaction.

Mixing and Stirring of the Raw Sludge With the Digested Sludge

Incoming fresh raw sludge should be thoroughly mixed with the digested sludge, by some effective method of agitation, so as to make a homogenous mass of raw as well as digested (or partly digested) sludge. In this way, the bacterial enzymes present in the digested sludge will get every opportunity to get mixed with the raw sludge, and to attack it for subsequent decomposition. The mixing of raw and digested sludge achieved by stirring the sludge in the sludge digestion tank by slow moving mechanical devices; or the gases of decomposition may be used to set up agitation by circulating from bottom to top of the tank and vice versa, by means of a pumping device. Excessive stirring may produce harmful effects, as it may kill the bacteria. The proper stirring however, results in even distribution of incoming sludge, breaks and reduces the scum, and helps in increasing the production of gases. In cold countries, where it is necessary to heat the digestion tanks, so as to maintain optimum mesophilic temperature (about 29°C), the stirring may help in transmitting heat from the heating coils to the tank contents; and thus to attain uniform temperature throughout the tank (NPCS Board of Consultants & Engineers, 2000).

SLUDGE DIGESTION TANK OR DIGESTERS

Aerobic Suspended Culture

A typical sludge digestion tank is shown in Figure 4. It consists of a circular RC.C tank with hoppered bottom, and having a fixed or a floating type of roof over its top. The raw sludge is pumped into the tank, and when the tank is first put into operation, it is seeded with the digested sludge from another tank, as pointed out earlier. A screw pump with an arrangement for circulating the sludge from bottom to top of the tank or vice versa (by reversing the direction of rotation of the screw) is commonly used, for stirring the sludge. Sometimes, power driven mechanical devices may be used for stirring the sludge, although these are not very popular, at present. In cold countries, the tank may have to be provided with heating coils through which hot water is circulated in order that the temperature inside the tank is maintained at optimum digestion temperature level.

The gases of decomposition (chiefly methane and carbon dioxide) are collected in a gas dome (in smaller tanks) or collected separately in gas holders (in larger tanks) for subsequent use. The digested sludge which settles down to the hoppered bottom of the tank is removed under hydrostatic pressure, periodically, once a week or so. The supernatant liquor lying between the sludge and the scum is removed at suitable elevations, through a number of withdrawal pipes, as shown. The supernatant liquor, being higher in BOD and suspended solids contents, is sent back for treatment along with the raw sewage in the treatment plant. The scum formed at the top surface of the supernatant liquor is broken by the re-circulating flow or through the mechanical rakers called scum breakers.

Design considerations

- Cylindrical in shape, circular in plan – diameter 3 to 12m
- Slope of bottom hopper floor – 1:1 to 1:3
- Depth of digestion tank – 6m
- Except in large plants not more than 2 units are provided.
- The capacity provided ranges from 21 to 61 lpcd
- If the progress of sludge digestion is assumed to be linear, then capacity of digestor (in cu.m) is

Figure 4. Cross section of a typical sludge digestion tanks

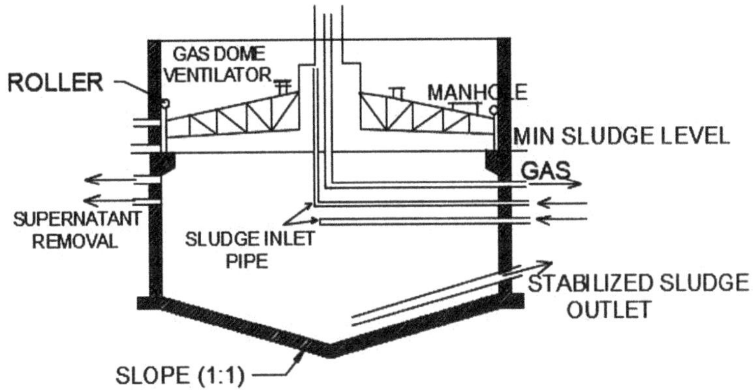

$$V = \left(\frac{V_1 + V_2}{2}\right)t \qquad (5)$$

where,

V_1 = raw sludge added per day, cu.m/d
V_2 = equivalent digested sludge produced per day, on completion of digestion
$V_2 = \left(\frac{V_1}{3}\right)$ t = digestion periods in days

When the daily digested sludge could not be removed, even though digestion gets completed, then consider separate capacity (Monsoon Storage). Thus total capacity,

$$V = \left(\frac{V_1 + V_2}{2}\right)t + V_{2.}T \qquad (6)$$

- **Realistic Case:** When the change during digestion is assumed to be parabolic then the

$$\text{average volume of digesting sludge} = [V_1 - \frac{2}{3}(V_1 - V_2)] \qquad (7)$$

Then,

$$\text{total capacity without monsoon storage} = [V_1 - \frac{2}{3}(V_1 - V_2)]t \qquad (8)$$

and

$$\text{total capacity with monsoon storage} = [V_1 - \frac{2}{3}(V_1 - V_2)]t + V_2T \qquad (9)$$

Two Stage High Rate Anaerobic Sludge Digestor

The treatment of wastewater sludge, from both primary and secondary treatment steps, consists of two main phases.

Two-stage digestion is carried out in a high-rate digester coupled in series with a second tank which is neither heated nor mixed. Its main function is to allow gravity concentration of digested solids and decanting of supernatant liquor. Decanting reduces volume of digested sludge, requiring further processing and disposal. The second tank fitted with a floating cover can also provide storage for digested sludge and digester gas. Very little solids reduction and gas production take place in the second tank. Many secondary digesters have performed poorly as

thickeners, producing dilute sludge and high-strength supernatant. This is because some gas will come out of the solution in small bubbles if there is incomplete digestion in the primary digester or if the sludge transferred from the primary digester is supersaturated with gas.

Figure 5. Two stage high rate anaerobic sludge digestor

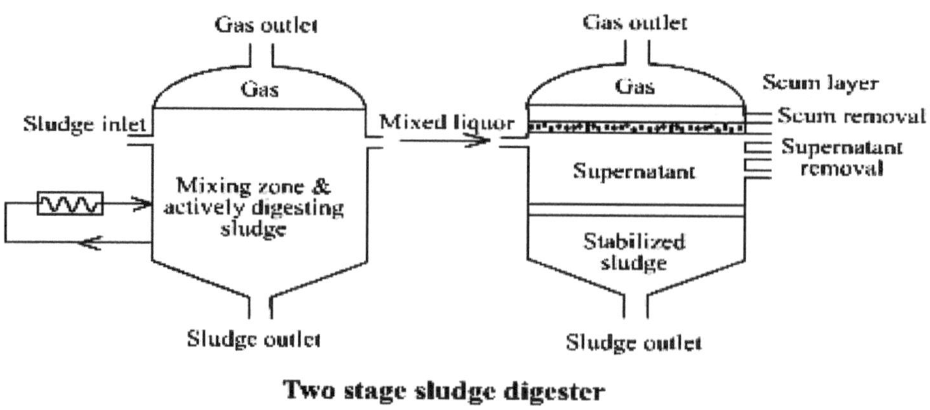

In two-phase digestion, acidogenesis & methanogenesis phases are divided into separate tanks coupled in series. The 1st reactor (acid-phase digester) is for hydrolysis and acidogenesis and is designed for 1-2 day SRT. It can be mesophilic or thermophilic. pH is between 5.5 and 6.5. Methane generation is negligible. Second reactor (methane-phase digester) is designed for about 10 days of SRT and operates in mesophilic temperature range.

Advantages of 2 phase digestion

1. Higher volatile solids reduction because it allows the creation of an optimum environment for the acid formers.
2. Increased production of gases
3. Higher content of methane in the final product gas
4. Higher pathogen reduction
5. Fewer foaming problems
6. Better stability of the digestion process

Dewatering and Drying of Digested Sludge

The digested sludge from the digestion tank contains a lot of water, and is, therefore, first of all, dewatered or dried up, before further disposal either by burning or dumping. These methods of dewatering the sludge are discussed below:

Dewatering, Drying and Disposal of Sludge by Sludge Drying Beds

Drying of the digested sludge on open beds of land (called sludge drying beds) is quite suitable for hot countries.

Design Considerations

1. Sludge Bed 15m x 30m in plan
2. Depth = 45 to 60 cm
3. Area of Sludge Drying Bed = 0.05 to 0.2sq.m per capita.
4. (d) Bottom layer - thick graded layers of gravel or crushed stone – size varying from 15cm at bottom to 1.25cm at top.
5. Top layer – 10 to 15cm thick coarse sand layer.
6. Open jointed under drained pipe – 15 cm diameter – 5 to 7 cm spacing – below gravel layer in valleys

The sewage sludge from the digestion tank is brought and spread over the top of the drying beds to a depth of about 20 to 30cm, through distribution troughs having openings of about 15 cm x 20 cm at a distances of about 2m or so. A portion of the moisture drains through the bed, while most of it is evaporated to the atmosphere. It usually takes about two weeks to two months, for drying the sludge, depending on the weather and condition of the bed.

Sludge should never be applied to a bed until the preceding dose has been removed. Hence, several drying beds will generally be required, with their number increasing with an increase in the number of days for which the sludge is kept on the beds. Normally, sludge is removed from the beds after a period of about 7 - 10 days; as within this period, about 30% of the moisture goes away and the surface of sludge gets cracked. The sludge cakes are then removed by spades, and they are dumped into a pit for further drying. The dried sludge is generally used as manure in our country, as it contains 1.7% nitrogen, 1.5% phosphoric acid, and 0.5% potash. It may also be used for filling up low lying areas. It may sometimes be disposed of by burning (i.e. by incineration).

Mechanical Methods of Dewatering Sludge

Dewatering is a physical (mechanical) unit operation used to reduce the moisture content of sludge so that it can be handled and/or processed as a semi-solid instead of as a liquid.

Devices commonly used for dewatering include: (i) Rotary Vacuum Filters (ii) Continuous Belt Filter Presses (CBFP) (iii) Filter Presses (iv) Centrifuges (v) ASP

Following is a detailed description of those dewatering devices considered capable of achieving the sludge disposal objectives.

Rotary Vacuum Filtration

Conventional

Rotary vacuum filtration basically consists of a cylindrical drum covered with a filter media which rotates partially submerged in a vat of sludge. The physical mechanisms which take place during vacuum filtration may be divided into three phases.

Figure 6. Schematic diagram of sludge drying beds

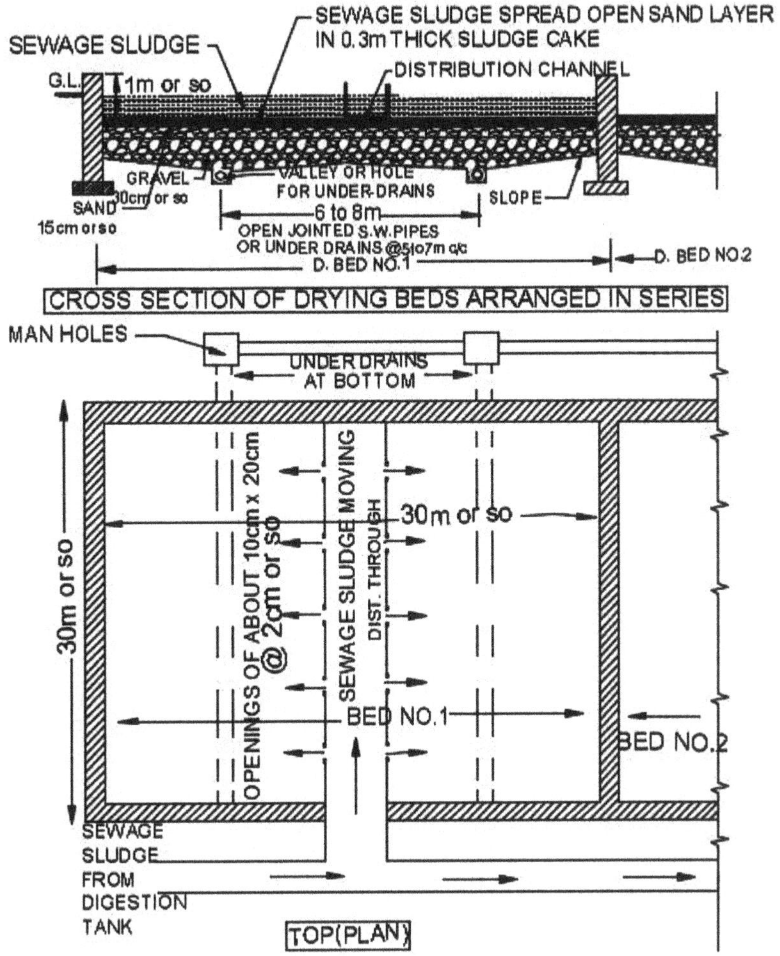

1. The first phase, which refers to the cake pick-up or form phase, occurs when a segment of the drum rotates into the sludge. Vacuum is applied to that segment, filtrate is drawn through the media and discharged. Concurrently, sludge solids are deposited on the media to form a partially dewatered cake. As the sludge cake increases in thickness, its resistance to the passage of filtrate *increases*.
2. The second phase, cake drying, occurs during that time the drum segment leaves the sludge and before the cake is removed. As the drum leaves the sludge, the cake is still under vacuum and additional moisture within the cake is drawn out.

Figure 7. Schematic diagram of Rotary Vacuum Filtration

3. The third phase, cake discharge, occurs after an acceptable cake dryness has been achieved and without vacuum. All of the above described operations are continuous in nature such that all three phases occur simultaneously on different portions of the drum. There are basically three types of rotary vacuum filters. These filters, described below, differ primarily in the type covering used and the cake discharge mechanism employed.

The drum filter was the original type unit employed in water and wastewater plants. Here, the filter media does not leave the drum which resulted in inadequate cake discharge and frequent belt washing. Use of this type unit has been virtually eliminated with the advent of the following two vacuum filtration units.

A coil vacuum filter uses two layers of stainless steel coils arranged in corduroy fashion around the drum. After the cake drying phase, the two layers of springs leave the drum and separate from each other. As a result of this separation, the cake is lifted off the lower layer of springs and discharged from the upper layer. The coils are then washed and reapplied to the drum. As would be expected, sludge with a large percentage of fine particles and resistant to flocculation, dewater poorly on coil filters.

The belt-type vacuum filter was introduced primarily to permit continuous washing of the filter media and thereby overcome plugging of the media by fines. As the filter media leaves the drum surface at the end of the cake drying phase it passes over a small discharge roll which facilitates cake discharge. These filters normally utilize a small diameter curved bar between the point where the belt leaves the drum and the discharge roll. This bar aids in maintaining belt stability and improves cake discharge.

Precoat

Precoat vacuum filtration is similar to conventional vacuum filtration with the exception of the application of a precoat prior to filtration. The precoat is normally diatomite, the siliceous skeletal remains of single cell aquatic plant life called diatoms. These diatoms form a permeable coating on the filter allowing filtrate to pass through easily while trapping sludge solids. Use of a precoat produces filtrate of very high quality.

Prior to filtration, the rotary drum is immersed in a slurry of precoat and an increasingly thick cake of diatomite is formed on the drum as the fluid is drawn through the media and solids deposited. After sufficient thickness is attained, the precoat cake is shaved smooth and the dewatering operation begun. Precoat vacuum filtration may also be divided into three phases with the first two being identical to those previously described. However, the third phase, cake discharge, begins when the cake has reached an acceptable dryness. The cake is usually removed before cracking occurs, as cracking tends to damage the precoat. During cake discharge, cake and a few thousandths of an inch of precoat are cut away by means of a continuously advancing knife. This fresh precoat surface is then rotated into the sludge to begin the dewatering cycle again.

The rotary vacuum filter for many years had been the standard used throughout the industry for sludge dewatering. While many of the earlier operational problems such as poor cake pick-up and release and high maintenance requirements have been somewhat improved in recent years by chemical conditioning methodology and mechanical innovations, two universal deterrents have prevented its continued widespread usage. These are:

1. The high energy and maintenance costs associated with operating a vacuum system.
2. The inability to produce as dry a cake as some of the more advanced technology currently available.

Vacuum Filter Operation

Filter cake formation is accomplished first by blinding of the media with the large particles and is followed by a packing of the pores near the filter media with the fine particles. Chemical conditioning of the sludge changes the size distribution of the sludge by coagulating the majority of small particles. This reduces resistance and improves filtrate clarity. Therefore, the size distribution of the sludge to be dewatered and the filter media specified have a significant impact on the extent to which a vacuum filter can dewater sludge.

Increased filter yield can be realized by increasing the vacuum due to the compressible nature of sludge. However, these benefits can be offset by higher power costs experienced in providing that vacuum. Increased drum submergence rate increases the time for cake pick-up and usually results in higher cake yields but at the expense of a wetter cake. Generally, drum submergence is kept between 15 and 25 percent to allow for adequate drying time to keep cake moisture content to a minimum.

Principal equipment includes a vacuum pump, filtrate receiver and pump, sludge pumping system, and in most applications, sludge conditioning apparatus. Each filter system is normally supplied with one vacuum pump and filtrate receiver. This receiver, which is interposed between the filter drum and vacuum pump, functions to separate air from the liquid and acts as a reservoir for the filtrate pump. Filtrate pumps must be sized to carry away that water separated out in the receiver.

Continuous Belt Filter Presses (CBFP)

First Generation CBFP

The continuous belt filter press was originally developed and in subsequent years, modified and improved in West Germany. Installation of the latest and best models in the United States has only recently experienced popularity. These systems were developed in an attempt to overcome the sludge pick-up problem occasionally experienced with rotary vacuum filtration. A combination of sludge conditioning, gravity dewatering and pressure dewatering is utilized to increase the solids content of the sludge.

With all units, the in feed sludge is mixed will polymer (or other chemicals) and placed onto a moving porous belt or screen. Dewatering occurs as the sludge moves through a series of rollers which squeeze the sludge to the belt or squeeze the sludge between two belts much like an old washing machine wringer. The cake formed is then discharged from the belt by a scraper

Figure 8. Schematic diagram of continuous belt filter presses

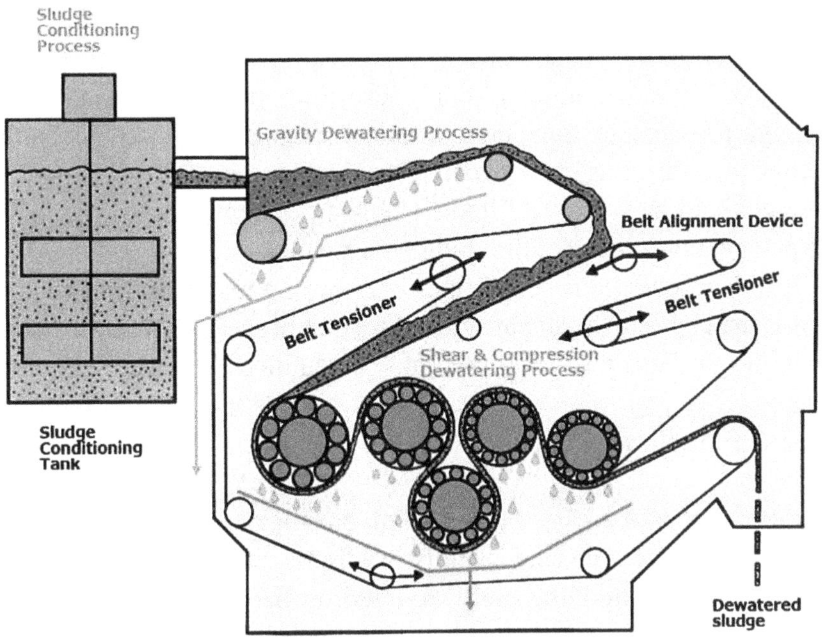

mechanism. There are basically three processing zones which occur along the length of the unit. These are: the initial drainage zone, which is analogous to the action of a drying bed; the press zone which involves application of pressure; and a shear zone in which shear is applied to the partially dewatered cake. Shearing action is accomplished by positioning the support rollers of the filter belt and the pressure rollers of the pressure belt in such a way that the belts and the sludge between them describe an S-shape curve. This condition creates a parallel displacement of the belts relative to each other due to the difference in radius.

Second Generation CBFP

In an attempt to improve the dewatering capabilities of the original CBFP's second generation was developed which incorporated a continuous mechanical thickening device in the initial stage. This thickening device is where the sludge is flocculated to promote rapid and complete cake formation. As a result, these units are capable of using a coarse mesh, relatively open weave, filter medium for improved gravity drainage.

Third Generation CBFP

Attempting to improve the dewatering capabilities of these units still further, a third generation of CBFP's have since been developed and are currently available from several manufacturers. These units employ additional medium and/or high pressure sections and variations in cake shearing mechanisms. These third generation units are capable of achieving dry cake solids equivalent to those attained by pressure filters. The use of chemicals to condition the sludge prior to dewatering of the sludge is a necessary step regardless of what type CBFP is selected. Proper sludge conditioning results in flocculation of the small particles into larger particles of sufficient size and strength to bridge the openings in the filter media and thus be retained on the belt.

Pressure Filtration

Of the several types of pressure filters available, the most widely used consists of a series of vertical plates held rigidly in a frame which are pressed together between a fixed and moving end. Mounted on the face of each individual plate is a filter cloth to support and contain the cake produced. Pressure filters do not produce a cake by pressing and squeezing. Instead, sludge is fed into the press "batch mode" through feed holes in trays along the length of the press. Pressures up to 225 PSI (16 kg/cm2) are applied to the sludge causing water to pass through the cloth while the solids are retained forming a cake

Figure 9. Schematic diagram of vertical pressure filtration

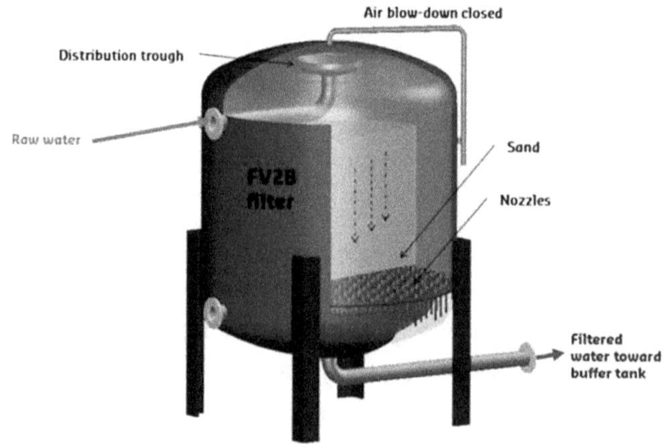

on the surface of the filter cloth. Sludge feed is stopped when the chambers between the trays are completely filled. Drainage ports are provided at the bottom of each chamber where the filtrate is collected, taken to the end of the press and discharged. The dewatering phase is complete when the flow of filtrate through the filter cloth nears zero. At this point, the sludge feed pump is stopped and any back pressure in the piping released. Each plate is then turned over the gap between the plates and the moving end to allow for cake removal. Filter cake usually drops below onto a conveyor for further removal.

After each plate has released its cake, the plates are pushed back together and the dewatering cycle restarted. In most applications, filter presses require a precoat material to aid in solids retention on the cloth and cake release.

Because of the pressures which may be applied by filter presses for the removal of water (5,000 to 20,000 times the force of gravity) cake solids produced are generally 10 to 20 percent drier, than those obtained by rotary vacuum filters under similar operating conditions. Recently, a pressure filter employing flexible rubber diaphragms between its chambers has been introduced. The feed slurry enters the top of the chamber between the filter cloths and gradually fills the chamber. After a cake is formed, the diaphragm is expanded by water under pressures to 250 lbs./in.2 (17.6 kg/cm2) which squeezes and dewaters the cake. The filter plates are then automatically opened and the cake discharged. Prior to the beginning of another pressing cycle, the filter cloths are washed. This press is reported to achieve shorter cycle times than conventional presses because of the improved control of the relationship between cake formation and pressure build-up.

However, pricing figures available indicate that these units will be priced about eight times the price of a conventional filter press. Widespread use of filter presses throughout the industry has not been experienced due to several operational limitations such as frequent filter cloth maintenance and manual operation of various dewatering sequences each contributing to excessive operational costs. In recent years, these shortcomings have been overcome with the advent of suitable monofilament filter media and increased mechanization.

The main advantage to using a filter press for sludge dewatering is in the reduced sludge disposal costs associated with producing drier cake solids. However, a detailed cost analysis should be performed to determine if these savings are sufficient to offset its high capital cost.

Centrifuge

A centrifuge is essentially a sedimentation device in which the solids - liquid separation is enhanced by the use of centrifugal force. This is accomplished by rotating the liquid at high speeds to subject the sludge to increased gravitational forces. There are basically three types of centrifuges available for sludge dewatering.

Continuous Solid Bowl Centrifuge

This centrifuge consists of two principal elements: a rotating bowl which is the settling vessel; and a conveyor which discharges the settled solids. The rotating bowl is supported between two sets of bearings and includes a conical section at one end. This section, which is not submerged, forms the dewatering beach or drainage deck. Sludge enters the rotating bowl through a stationary feed pipe extending into the hollow shaft of the rotating screw conveyor and is distributed through ports into a pool within the bowl. As the bowl rotates, centrifugal force causes the slurry to form an annular pool, the depth of which is determined by the effluent fluent weirs. The rotating screw conveyor continuously moves the sludge solids across the bowl, up the beaching incline to outlet ports and then to a discharge hopper for ultimate removal. As the liquid sludge flows through the bowl towards the overflow weirs, progressively finer solids are forced to the rotating bowls wall.

Figure 10. Schematic diagram of centrifuge

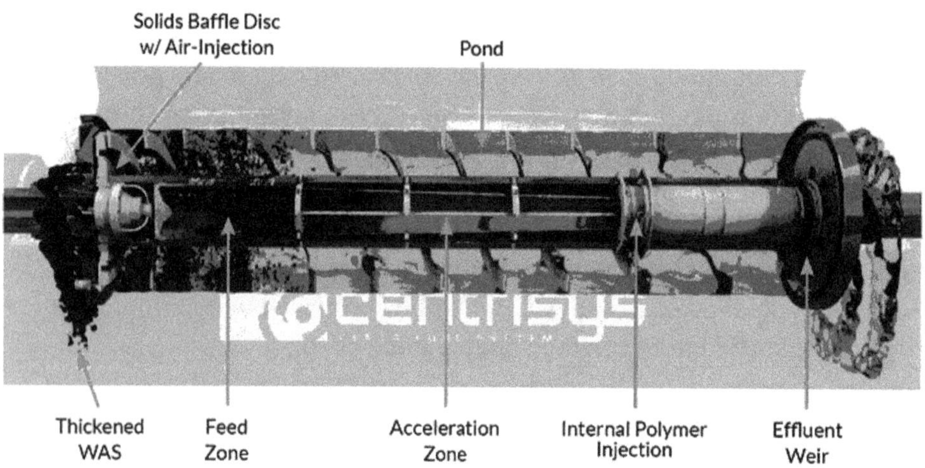

The water (centrate) drains from the solids and back into the pool. The centrate is then discharged through ports at the end where the effluent weirs are located. Most solid bowl centrifuges are of the "counter current" type employing a counter current flow of liquid and solids. Recently a "concurrent" centrifuge has been introduced in which the incoming sludge is carried by the feed pipe to the end of the bowl opposite the discharge. As a result, settled solids are not disturbed by the incoming feed. Turbulence is reduced substantially as both the solids and liquid pass through the bowl in smooth parallel flow patterns. Because solids are conveyed over the entire length of the bowl before discharge, better compaction is achieved providing drier cake solids and reduced polymer demand.

Basket Centrifuge

The basket or imperforated bowl-knife discharge unit, is a batch dewatering unit introduced primarily for use as a partial dewatering device for small operations. Sludge is charged into the basket forming an annular ring as the unit rotates around its vertical axis. Cake continually builds up within the basket as the liquid (centrate) is displaced over a baffle or weir at the top of the unit. When the solids concentration of the centrate reaches an undesirable level, the centrifuge is stopped, the unit decelerates, and a skimmer enters the bowl to remove the remaining liquid. A knife is then moved into the bowl to cut out the cake which falls out the open bottom for removal. Although the batch operating mode of this type unit does limit its adaptability to large scale operations, it is capable of producing higher solids recovery than continuous devices without chemical addition because of minimum disturbance of the depositing solids.

Disc Centrifuge

The disc centrifuge, is a continuous flow variation of the previously described basket centrifuge. The incoming sludge is distributed between multitudes of narrow channels formed by stacked conical discs. Sludge solids have only a short distance to settle which rather limits this units application to those sludges containing small, low density particles. Solids are collected and continuously discharged through fairly small orifices in the bowl wall (0.05 to 0.10 inches). These orifice openings impose the upper limit as to the size solids which can be handled by this type unit. Without adequate screening prior to centrifugation, this centrifuge is prone to orifice plugging resulting in prohibitive maintenance costs.

The dewatered sludgc obtained from mechanical devices of western countries, is generally heat dried, so as to produce fertilizers, As a matter of fact, the mechanical dewatering removes only about 50% of the moisture, and hence the mechanically dewatered sludge is actually heated, so as to fully remove the moisture from it. The dry residue is used as manure. This method is looked upon more as a method of producing fertilizer rather than as a method of sludge disposal, because if this method is adopted only for sludge disposal, it proves to be extremely costly, and thus feasible only for rich countries. The wet sludge, after mechanical dewatering, is sometimes, directly disposed of either in sea or in underground trenches, or burnt, as discussed below:

DISPOSAL OF DIGESTED SLUDGE

Disposal by Dumping Into the Ocean

Coastal cities have discharged digested sludge into the ocean for half a century. Just like crayfish have got used to be dumped into boiling water through the centuries, the oceans may have adapted to this sludge dumping. Dewatered wastes may be transported to offshore sites in barges and dumped, or sludge slurry may be dumped to deep water through a submarine outfall. In recent years, this practice has been questioned by regulatory agencies. The principal environmental concerns are degradation of recreational waters, build-up of solids on the sea bottom, and toxicity to marine life. The contaminants involved are the same as those related to disposal of land - heavy metals, pathogens, and organic pollutants.

Ocean disposal sites should have adequate current velocities for initial dilution and waste dispersion to prevent concentration of pollutants. Some waste slurries have a lower specific gravity than seawater and rise to the surface, while the solids of others accumulate on the bottom creating sludge deposits that decompose. The content of heavy metals in municipal sludges can be limited by instituting and enforcing a sewer ordinance to control industrial wastewater quality. As with land disposal, the sludge should be biologically stabilized to reduce pathogens.

A dumpsite in the Atlantic Ocean used by the city of Philadelphia to discharge digested sludge has been monitored for several years with little or no evidence of harmful effects found in the marine ecosystem. For example, faecal coliform levels were considered safe for shellfish harvesting and metal accumulations in the sediments did not show an obvious pattern. Another extensive study was conducted in coastal waters off southern California where several cities discharge wastewaters and the Hyperion plant discharges digested sludge through a submarine outfall. Except for localized areas near the ends of the outfall pipes, environmental conditions were not measurably influenced, and within these areas the major evidence of pollution was an increase in the organic matter on the bottom. The investigation was extensive, incorporating the sources and distribution of pollutants. For coastal cities, this method of ultimate disposal should be considered as one option along with incineration, land application, and permanent storage in landfill.

Disposal by Burial Into the Trenches

This method is used principally for raw sludge, where, unless covered by earth, serious odor nuisances are created. The sludge is run into trenches two to three feet wide and about two feet deep. The raw sludge in the trenches should be covered by at least 12 inches of earth. Where large areas of land are available, burial of raw sludge is probably the most economical method of sludge disposal as it eliminates the costs of all sludge treatment processes. It is, however, rarely used and even then as a temporary makeshift because of the land area required. The sludge in the trenches may remain moist and malodorous for years so that an area once used cannot be reused for the same purpose or for any other purpose for a long period of time.

Disposal by Incineration

The dewatered wet sludge produced in waste water treatment plant may also be disposed of by burning, in suitably designed incinerators, when sufficient space is not available for its burial near the plant site, or the sludge cannot be dried and used as manure. The following types of incinerators (furnaces) have primarily been designed and used for incinerating wet sludge:

1. Multiple-hearth furnace
2. Fluid-bed furnace
3. Flash-type of furnace
4. Infrared (Electric or Radiant heat) furnace.

Use of Lagoons for Disposal of Raw Sludge

This method is, sometimes, used at smaller places for disposing of raw sludge without digestion. In this method, the raw sludge is kept at rest in a large shallow open pond, called a lagoon. The detention period is 1 to 2 months, and may extend up to 6 months. During its detention in the lagoon, the sludge undergoes anaerobic digestion thereby getting stabilized. Due to this anaerobic decomposition of sludge, foul gases will be evolved from a lagoon; and hence the lagoons should be located away from the town, and direction of the common winds should be such that the smells are not carried towards any localities.

A typical section for a lagoon is shown in Figure10 . It is a shallow pit, 0.6 to 1.2m deep, formed by excavating the ground.

At the bottom of this pit, a 15 cm thick layer of ashes or clinker is placed. Agricultural tile drains of about 10 cm diameter are laid at bottom as under-drains. These are placed at about 3m centre to centre spacing. Banks are formed on both sides of the pit from the excavated earth as shown.

After the sludge has been stabilized, and the moisture has been drained away or evaporated during its detention-fin the lagoon, the contents are dug out to about half of their original volume, and used as manure. This method of sludge disposal is quite cheap (as no digestion tanks are required), but the greatest drawback is the evolution and eruption of foul gases, polluting the environment. Its use is, therefore, restricted only to non residential areas.

Figure 11. Typical section of a Lagoon for disposal of raw sludge

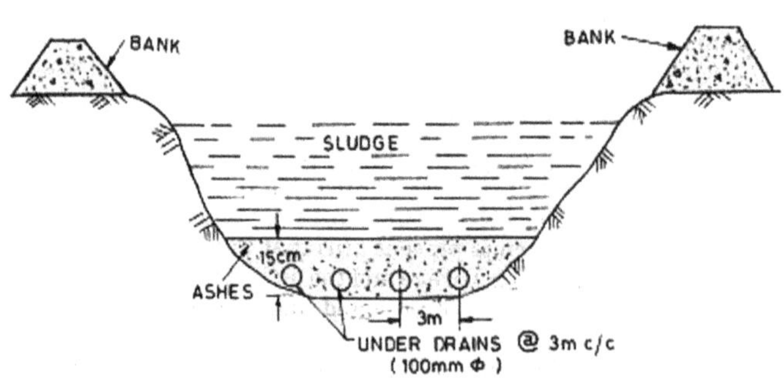

Sludge Stabilization

Sludge is stabilised to reduce pathogens, to eliminate offensive odours and to inhibit, reduce or eliminate the potential for putrefaction. The success in achieving these objectives is related to a reduction of the organic (volatile) fraction or the addition of chemicals to the sludge to render it unsuitable for the survival of microorganisms. Stabilization is usually accomplished by biological or chemical treatment processes.In addition to the health an aesthetic reasons mentioned above, stabilisation is used for volume reduction, production of usable gas (methane), and improving the dewaterability of sludge.

Chemical Stabilization

Chemical stabilization of sludges is aimed not at reducing the quantity of biodegradable organic matter, but at creating conditions that inhibit microorganisms in order to retard the degradation of organic materials and prevent odors. The most common chemical stabilization procedure is to raise the pH of sludge using lime or other alkaline material, such as cement kiln dust. Sludge can be chemically stabilized in liquid or dewatered forms. When dewatered sludge is used, the exothermic reaction of lime with water causes heating which helps destroy pathogens and evaporate water.

Biological Stabilization

In biological stabilization processes, the organic content of sludges is reduced by bio-logical degradation in controlled, engineered processes. Most commonly, domestic wastewater sludge is biologically stabilized as a liquid in anaerobic digesters from which methane gas is a byproduct. Liquid sludge can also be biologically stabilized in aerobic digesters to which oxygen (or air) must be added. Composting is a process that biologically stabilizes dewatered sludge. Composting is ordinarily an aerobic process, and an amendment such as wood chips or sawdust must be added to improve friability in order to promote aeration. Composting takes place at thermophilic temperatures (often, about 55°C) because of heat released by biochemical transformations. Aerobic digesters can be made to operate thermophilically using heat from the same source. Anaerobic digesters can operate at thermophilic temperatures by burning methane produced from the process, but they typically operate at mesophilic temperatures (at about 35°C).

NUMERICAL PROBLEMS

Example 1

Design a digestion tank for the primary sludge with the help of following data:

1. Average flow = 25MLD;
2. Total suspended solids in raw sewage = 350mg/l;
3. Moisture content of digested sludge = 85%

Assume any other suitable data you require
Sol. Average sewage flow = 25MLD
Total suspended solids = 350 mg/l
Therefore, mass of suspended solids in 25 million litre of sewage flowing per day

$$= \frac{350 \times 25 \times 10^6}{10^6} \text{kg} = 8750\text{kg/day}$$

Assuming that 70% solids are removed in primary settling tanks, we have
Mass of solids removed in the primary settling tank

$$= 70\% \times 8750\text{kg/day} = 6125 \text{ kg/day}$$

Assuming that the fresh sludge has a moisture content of 95%, we have 5kg of dry solids will make = 100 kg of wet sludge
Therefore 6125kg of dry solids will make

$$= \frac{100}{5} \times 6125 \text{ kg of wet sludge per day}$$

$$= 122{,}500 \text{ kg of wet sludge per day}$$

Assuming the sp. gravity of wet sludge as 1.02, we have
The volume of raw sludge produced/day

$$= V_1 = \frac{122{,}500}{1020} \text{m}^3\text{/day} = 120.09 \text{ m}^3\text{/day}$$

The volume of the digested sludge (V_2) at 85% moisture content is given by the formula as

$$V_2 = V_1 \times \frac{100 - P_1}{100 - P_2} \tag{10}$$

$$V_2 = V_1 \times \frac{100 - 95}{100 - 85} \text{ or } V_2 = V_1 \times \frac{5}{15}$$

$$V_2 = V_1 \times \frac{5}{15} = 120.09 \times \frac{1}{3} \text{ m}^3\text{/day} = 40.03\text{m}^3\text{/day}$$

Now assuming the digestion period as 30 days, we have the capacity of the required digestion tank, given by:

$$\text{Capacity} = [V_1 - \frac{2}{3}(V_1 - V_2)] \times t \tag{11}$$

$$= [120.09 - \frac{2}{3}(120.09 - 40.03)] \times 30 = 2001.5\text{m}^3$$

Now, providing 7m depth of the cylindrical digestion tank, we have

$$\text{Cross-sectional area of the tank} = \frac{2001.5}{7} = 285.92\text{m}^2$$

$$\text{Therefore diameter of the tank} = \sqrt{\frac{285.92}{\pi / 4}} \text{ m} = 19.07\text{m}$$

Hence, provide a cylindrical sludge digestion tank 7m deep and 19.07m diameter, with additional hoppered bottom of 1:1 slope for collection of digested sludge.

Example 2

A sewage containing 250mg/l of suspended solids is passed through primary settling tank. The solids from the primary settling tank are digested to recover the gas. Find the likely volumes of methane and carbon dioxide present in the digestion of the sludge from 11000m^3 of sewage. Calculate the fuel value of the gas produced. State clearly the assumptions made.

Sol. Total suspended solids in sewage = 250mg/l

Assuming that 60% of suspended solids are removed in the primary settling tank, we have

The suspended solids removed as sludge = 60% × 250mg/l

=150mg/l

Now, assuming that the volatile solids present are 70% of the suspended solids we have,

Therefore the volatile solids removed =70% ×150mg/l =105mg/l

Further assuming that the volatile matter is reduced by 65% in sludge digestion, we have

Volatile matter reduced = 65% × 105mg/l = 68.25mg/l

Hence, volatile matter reduced in 10000cu.m. of sewage

$$= 68.25 \times \frac{10,000 \times 1000}{10^6} \text{kg} = 682.5\text{kg}$$

Now, assuming that 0.9cu.m. of gas is produced per kg of volatile matter reduced, we have

Total quantity of gas produced = 0.9× 682.5cu.m = 614.25 cu.m

Assuming that the produced gas contains 68% methane and 30% carbon dioxide, we have

Methane produced = 0.68×614.25cu.m = 417.69cu.m

Carbon Dioxide produced = 0.30×614.25 cu.m =184.27cu.m

Now assuming that the methane in the sludge gas has a fuel value of 38000Kj/m^3, we have

The fuel value = 36000×417.69 kJ = 15.03MkJ

Now assuming a boiler efficiency of 80%, we have the amount of heat that can be

furnished by the boiler = 80%×15.03MkJ = 12.024MkJ

Example 3

Design sludge drying beds to dewater the digested sludge produced from wastewater treatment plant based on the activated sludge process designed for 50000 population. Assume other suitable data.

1. Sol. Compute the amount of dry solids produced per day

Assuming dry solids concentration in digested primary and activated mix sludge to be 70g/capita/d,

Dry solids produced = 50000 (capita) × 70(g/capita/d)

= 3500000g/d

= 3500kg/d

2. Compute the area of drying beds needed

Assuming dry solids loading rate of 100 kg/m^3/year,

Area of bed needed = $\frac{dry\, solids\, applied \,/\, year}{dry\, solid\, loading\, rate}$

$$= \frac{3500 \times 365 \left(\frac{kg}{year}\right)}{100 kg \,/\, m^2 \Big/ year} = 12775 \text{ m}^2$$

3. Check for the area of drying bed per capita

Bed area per capita = 12775/60000 = 0.255 m^2/capita

It is acceptable as within an acceptable range of 0.10 to 0.25 m^2/capita

4. Compute the number of beds

Providing 30m long × 8m wide beds,

The number of beds needed $= \frac{total\, area\, of\, the\, beds}{area\, of\, one\, bed} = \frac{12775\text{m}^2}{30(m) \times 8(m)}$

$\approx 54 m$

5. Compute the depth of sludge applied on each bed

Assuming 7% solids and specific gravity of mixed digested sludge as 1.02,

Volume of digested sludge $= \frac{W_s}{\rho_w \times S_{sl} \times P_s} = \frac{3500\left(\frac{kg}{d}\right)}{1000\left(\frac{kg}{m^3}\right) \times 1.02 \times 0.07} =$

49m^3/d

Therefore,
the sludge depth

$$= \frac{total\,volume\,of\,sludge\,applied\,/\,year}{number\,of\,beds \times area\,of\,each\,bed \times number\,of\,cycles\,of\,each\,bed}$$

$$= \frac{49\ (m^3\,/\,d) \times 365\left(d\right)}{54 \times 240\left(m^2\right) \times 10} \approx 15cm$$

Therefore, provide 56beds (54 working + 2 standby), each of 30m×8m, with the standard sand and gravel media and under drainage system.

Check for the number of beds: The volume of sludge generated per annum is

49×365 m^3/year = 17885 m^3/year

The name of sludge that each bed, with 10 cycles of operation per annum, can handle is given by

= number of cycles × area of bed × depth of sludge on the bed

= 10 × 30m × 8m × 0.15m = 360 m^3/year

Therefore,
the number of beds required

$$= \frac{total\,volume\,of\,sludge\,per\,annum}{volume\,of\,sludge\,each\,bed\,can\,handle\,per\,annum}$$

$$= \frac{17885\,m^3\,/\,year}{360\,m^3\,/\,year} = 50\text{ beds}$$

REFERENCES

Andreoli, C. V., Sperling, M. V., & Fernandes, F. (2007). *Sludge Treatment and Disposal* (Vol. 6). IWA Publishing.

Davis, M., & Cornwell, D. (n.d.). *Introduction to Environmental Engineering* (3rd ed.). McGraw-Hill.

Garg, N. K. (2009). A Thesis on Multicriteria Assessment of Alternative Sludge Disposal Methods. In *Investigation of inorganic materials derived from water purification processes for ceramic applications*. Water Research Commission Report No 538/1/95, WRC.

Garg, S. K. (2015). *Sewage Disposal and Air Pollution Engineering* (Vol. 2). Khanna Publishers.

Gerardi, M. (2003). *The microbiology of anaerobic digestors*. John Wiley & Sons Inc.

Hope, J. (1986). *Sewage Sludge Disposal and Utilization Study Report*. Academic Press.

Metcalf & Eddy. (2003). *Wastewater Engineering, Treatment and Reuse* (4th ed.). Tata McGraw Hill Publishing Co. Ltd.

NPCS Board of Consultants & Engineers. (2000). *Water and Air Effluents Treatment Handbook: Air Pollution Control Equipment*. Author.

Sharma, B. K. (2007). *Environmental Chemistry*. Krishna Prakashan Media Ltd.

Turovskiy, I. S., & Mathai, P. K. (2006). *Wastewater Sludge Processing*. John Wiley & Sons. doi:10.1002/047179161X

Tyagi, V. K., & Lo, S. L. (2016). Energy and Resource Recovery From Sludge: Full-Scale Experiences. In *Environmental Materials and Waste Resource Recovery and Pollution Prevention*. Elsevier.

Chapter 5
Advanced Wastewater Treatments

ABSTRACT

Advanced wastewater treatment is the process that reduces the level of impurities in wastewater below that attainable through conventional secondary or biological treatment. It includes the removal of nutrients such as phosphorus and nitrogen and a high percentage of suspended solids. The removal of nitrogen and phosphorus from wastewater has become an emerging worldwide concern because these compounds cause eutrophication in natural water. A post-treatment process is therefore required to remove nitrogen and phosphorus from the effluent. Therefore, the purpose of this chapter is to provide the deeper knowledge of membrane technology, membrane bioreactor, sequential batch reactor, moving bed biofilm reactor, nitrification, denitrification, phosphorus removal from wastewater, carbon adsorption, and provide a design of a sewage treatment plant using moving bed biofilm reactor technology.

MEMBRANE TECHNOLOGY

Membrane technology is easy and well-arranged process conductions. The membrane acts as a very precise filter that stops suspended solids and other substances to pass through, while it allows water to pass through. There are two factors that determine the effectiveness of a membrane filtration processes: retention and flux. Membrane filtration can be divided into micro

DOI: 10.4018/978-1-5225-9441-3.ch005

and ultrafiltration (MF/UF), Nanofiltration (NF) and reverse osmosis (RO). When membrane filtration is used for the removal of larger particles MF/UF are applied. The pressure that is required to perform NF and RO is much higher than the pressure required for MF/UF, while productivity is much lower (Sonune et al., 2004).

A recent advancement in wastewater treatment technology involves the filtration of wastewater through porous membranes (Sharrer et al., 2007). These technologies have introduced a new cutting edge on wastewater treatment. For concentrated wastewaters, like industrial streams and landfill leachate, MBR has been applied at full scale successfully however this system requires relatively high energy. Using new membrane techniques, like transfer flow modules, creates the possibilities of a more widespread application. MBR technologies provide the potential for reuse wastewater generated from industries or municipalities and decrease in sludge production. The MBR combines suspended growth unit responsible for the biodegradation of the waste compounds and the membrane filtration module for the physical separation of the treated water from the mixed liquor using a porous membrane that helps to retain high microbial concentration in the reactor and increase the biological operation capacity of the reactor. The MBR process was introduced by the late 1960s, as soon as commercial scale UF and MF membranes were available (Le-clech et al., 2006). The original process was introduced by Dorr-Olivier Inc. and combined the use of an activated sludge bioreactor with a cross flow membrane filtration loop (Smith et al., 1969).

Although the research on MBR technology began only few decades ago, it has developed quite rapidly and become one of the important technologies in wastewater treatment process. Up to this date, MBR systems have mostly been used to treat industrial wastewater, domestic wastewater and specific municipal wastewater. Requirement of higher removal of organic matters, suspended solids, nutrient and harmful bacteria from the wastewater and the requirement to meet the strict effluent discharge quality in terms of nutrient and micro-pollutants, the main cause for the eutrophication and decrease the water quality in the receiving water bodies, are the important issues in the present wastewater treatment processes (Ersu et al., 2008; Kraume et al., 2005). MBR technology have become a most promising process to overcome these issues and the nutrient removal from the wastewater and several studies have been focused on nutrient removal from wastewater using MBR (Galil et al., 2009; Ersu et al., 2008; Yuan et al., 2008; Kraume et al., 2005; Song et al., 2004; Adam et al., 2002; Lesjean et al., 2002).

This process have several advantages i.e. higher treatment efficiency is obtained in smaller footprints compared to conventional treatment processes, higher biomass concentrations, no long sludge-settling periods, lower sensitivity to toxic compounds and both organic and high ammonia removals in a single process, flexible in terms of shape, load and volume, minimize labour costs, no need to add any chemicals. The disadvantages are high operating costs associated with the aeration process, membrane replacement costs are high and must be budgeted for appropriately, concentrate and waste stream disposal issue (Sonune et al., 2004).

Membrane Bioreactor

MBR systems have mostly been used to treat industrial wastewater, domestic wastewater and specific municipal wastewater, where a small footprint, water reuse, or stringent discharge standards were required. It is expected that the MBR system will increase in capacity and broaden in application area due to future, more stringent regulations and water reuse initiatives. The advantages of this process have higher and more consistent effluent quality can be achieved even in smaller footprints and smaller reactor volume, less dependent on mixed liquor suspended solid (MLSS) concentration and sludge volume index (SVI), less sludge production, no need of operators can operate automatically. The disadvantages of the membrane bioreactor process i.e. relatively expensive to install and operate, required frequent monitoring and maintenance, limitations imposed by pH, temperature and pressure requirements to meet the membrane tolerance, les efficient oxygen transfer due to high MLSS concentration.

Sequencing Batch Reactor (SBR)

Sequencing batch reactors are a type of activated sludge process for the treatment of wastewater. SBR reactors treat wastewater such as sewage or output from anaerobic digesters or mechanical biological treatment facilities in batches. Oxygen is bubbled through the mixture of wastewater and activated sludge to reduce the organic matter (measured as biochemical oxygen demand (BOD) and chemical oxygen demand (COD)). The treated effluent may be suitable for discharge to surface waters or possibly for use on land.

There are five stages in treatment process:

Figure 1. Schematic diagram of conventional activated sludge process (top) and external membrane bioreactor (bottom)

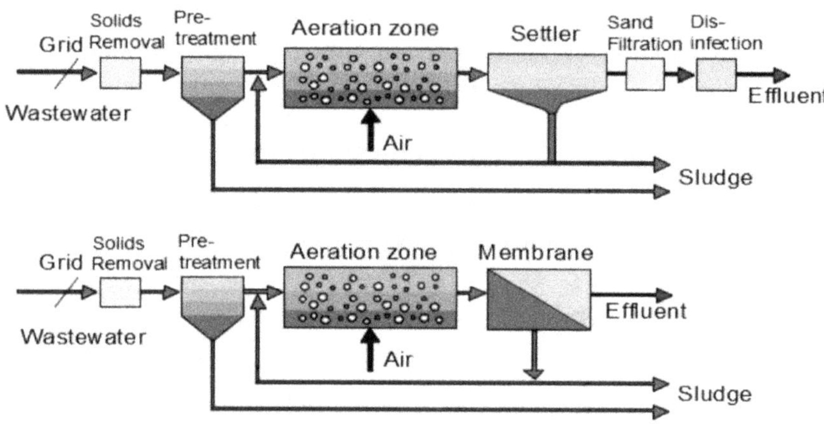

1. **Fill:** The influent to the tank may be either raw wastewater (screened and degritted) or primary effluent. It may be either pumped in or allowed to flow in by gravity. The feed volume is determined based on a number of factors including desired loading and detention time and expected settling characteristics of the organisms. The time of Fill depends upon the volume of each tank, the number of parallel tanks in operation, and the extent of diurnal variations in the wastewater flow rate. Virtually any aeration system (e.g., diffused, floating mechanical, or jet) can be used. The ideal aeration system, however, must be able to provide both a range of mixing intensities, from zero to complete agitation, and the flexibility of mixing without aeration. Level sensing devices, or timers, or in-tank probes (e.g., for the measurement of either dissolved oxygen or ammonia nitrogen) can be used to switch the aerators and/or mixers on and off as desired.
2. **React:** Biological reactions, which were initiated during Fill, are completed during React. As in Fill, alternating conditions of low dissolved oxygen concentrations (e.g., Mixed React) and high dissolved oxygen concentrations (e.g. Aerated React) may be required. The liquid level remains at the maximum throughout react, sludge wasting can take place during this period as a simple means for controlling the sludge age. By wasting during react, sludge is removed from the reactor as a means of maintaining or decreasing the volume of sludge in the reactor and decreases the solids volume. Time dedicated to react can be as high as

50% or more of total cycle time. The end of react may be dictated by a time specification (e.g. the time in React shall always be 1.5 h) or a level controller in an adjacent tank.

3. **Settle:** In the SBR, solids separation takes place under quiescent conditions (i.e., without inflow or outflow) in a tank, which may have a volume more than ten times that of the secondary clarifier used for conventional continuous-flow activated sludge plant. This major advantage in the clarification process results from the fact that the entire aeration tank serves as the clarifier during the period when no flow enters the tank. Because all of the biomass remains in the tank until some fraction must be wasted, there is no need for underflow hardware normally found in conventional clarifiers. By way of contrast, mixed liquor is continuously removed from continuous-flow activated-sludge aeration tank and passed through the clarifiers only to have major portion of the sludge returned to the aeration tank.
4. **Draw (Decant):** The withdrawal mechanism may take one of several forms, including a pipe fixed at some predetermined level with the flow regulated by an automatic valve or a pump, or an adjustable or floating weir at or just beneath the liquid surface. In any case, the withdrawal mechanism should be designed and operated in a manner the prevents floating matter from being discharged. The time dedicated to draw can range from 5 to more than 30% of the total cycle time. The time in draw, however, should not be overly extended because of possible problems with rising sludge.
5. **Idle:** The period between draw and Fill is termed Idle. Despite its name, this “idle” time can be used effectively to waste settled sludge. While sludge wasting can be as infrequent as once every 2 to 3 months, more frequent sludge wasting programs are recommended to maintain process efficiency and sludge settling.

The advantages of the SBR over conventional system can be summarized as follows:

1. Control system provided high flexibility: the control system automatically coordinates equipment operation through various phases of SBR cycle. This feature offers a high degree of flexibility allowing adaptation of the process cycle to meet the changing influent conditions through simple changes in control set points.

Figure 2. Flow chart of sequential batch reactor

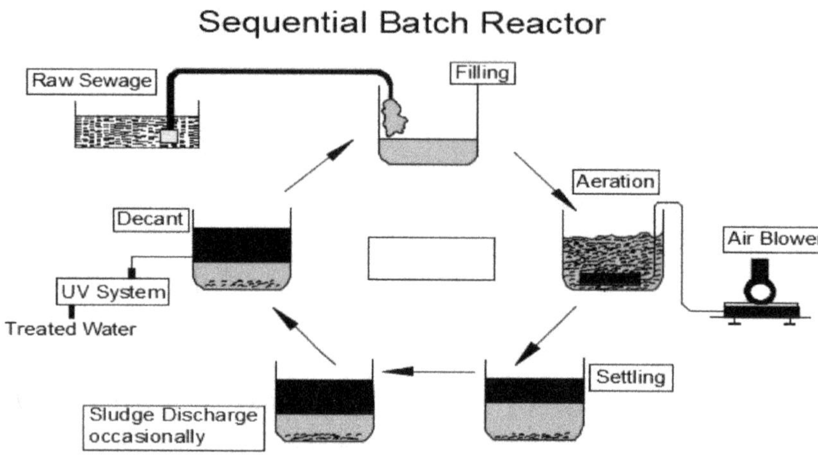

2. No primary and secondary settling tanks, no return sludge pumping, hence lesser area requirement and ease in operation and maintenance.
3. Proven process, which enhances the standard system through strategic cost, operating and biological advantages.
4. Improved influent quality: Extended aeration mode, a special ability to handle extremely high organic and hydraulic shock loads, no washout of biomass, reliable performance. More than 95% BOD removal, advantages same as extended aeration processes.
5. Proven effluent quality below 10 mg/L BOD_5 and TSS
6. Proven nutrient removal quality below 1 mg/L ammonia.
7. Improved sludge settling: Due to pre-react zone. Increased settling area without sludge scrapers provide perfectly quiescent settling environment. There is no short circuiting due to density current, temperature difference and uneven inflow distribution as in settling tanks.
8. Low volume of sludge production. Lower annual operating costs due to low volume sludge pumping, low waste sludge production, savings in precipitating chemicals and savings for cost of operating personnel and equipment.

Moving Bed Biofilm Reactor (MBBR)

The MBBR technology was developed in Norway in the late 1980's and early 1990's when the nitrogen removal from the wastewater was the main focus and later on organic matter removal has been more investigated (Odegaard, 1999). This technology adopts the best from activated sludge processes and biofilter processes and can be operated as a standalone process or it can be used to specifically enhance or upgrade the treatment capacity of old plants which has limited space for the future extension. This process has become popular in the field of wastewater treatment because it maximizes the capacity and efficiency of the treatment plant while minimizing the footprints. It has the capacity to withstand the challenges of wastewater industry like; retrofitting the old treatment plants, higher nutrient removal capacity, produce less sludge as a result of high biomass retention time, minimize process complexities and operators, no need of backwashing, easy maintenance, economical, self regulating process with fluctuating organic loads and so on. MBBR systems are mainly based on the aeration rate and reactors filled with the specially designed carriers to provide a surface to colonize by bacteria (Rahimi et al., 2011). When the suspended porous biofilm carriers are kept in continuously mixed and operated aeration tank, active biomass grows as a biofilm on the surface of these carriers having a density slightly less than the water (Kermani et al., 2008). MBBR system is the efficient method to retain slow growing microorganisms such as nitrifiers in the form of biofilm.

The MBBR has become popular on the broad range of wastewater treatment as an enhancement of biological nutrient removal. MBBR is a continuously operating, non-cloggable biofilm reactor with low head loss and high specific biofilm. For this process, specially designed biofilm carrier is required in which microorganisms start growing while moving continuously with water in the reactor. This process improves reliability, simplifies operation and requires less space as compared to other conventional treatment processes.

The MBBR system can be operated under aerobic conditions for BOD removal and nitrification or under anoxic conditions for denitrification. During operation, the carriers are kept in constant circulation. In an aerobic reactor, circulation is induced through the action of air bubbles injected into the tank by a coarse bubble diffuser system. In an anoxic reactor, a submerged mechanical mixer is typically supplied. Specific area of the biomass carrier, flow and the mixing condition in the reactor and DO concentration are the main factors that affect the operation of MBBR. The advantages of this treatment processes, it required small footprint and reactor volume, high effluent quality in terms of nutrient removal, good disinfection capability, higher volumetric loading, shock load protection and less sludge production, ease in upgrade of existing facilities. The disadvantages which are incurred in this process are high equipment and operation cost, fouling or biofouling of the membrane due to deposits of inorganic, organic & microbiological materials on the membrane surface and inside the pores. Extensive fouling leads to a pronounced decrease in permeate flux and can threaten the economic efficiency of the membrane plant (Shore et al., 2012).

Different Types of Media Used in MBBR

The media on which the biofilm develops are carefully designed with high internal surface area having density slightly less than the water so that it can easily float. The most commonly used solid surface for attached growth processes are; stones, clinker, sand, activated charcoal, metals, plastic sheets, and foams. There are different types of media which can be used as a media for the microbial growth. The physical appearance and characteristics of these media are shown in figure. The biofilm carrier is selected to have low density close to water (sponge or plastic carriers), high specific surface area, good holding capacity, and it must avoid the clogging by increased biomass. By means of biomass carriers, it is possible to obtain a two fold increase in biomass concentration in the aeration tanks compared to that in the conventional activated sludge process (Jianlong et al., 2000). Tavares et al. (1995) stated that the microorganisms produce a kind of natural polymer which helps them to attach to the surface of inert carrier resulting the biofilm layer formation.

DESIGNING OF TREATMENT PROCESS

STP Designing Using MBBR Technology

Parameters

Assume the flow is 1500 kld, BOD: 300 ppm, TSS: 250 ppm

The average flow can be calculated by the given formula below:

Average Flow

$$\frac{Flow(kld)}{24(hours)} = \frac{1500}{24} = 62.5 \text{ m}^3\text{/hr}$$

Figure 3. Physical appearance of the media used in attached growth processes

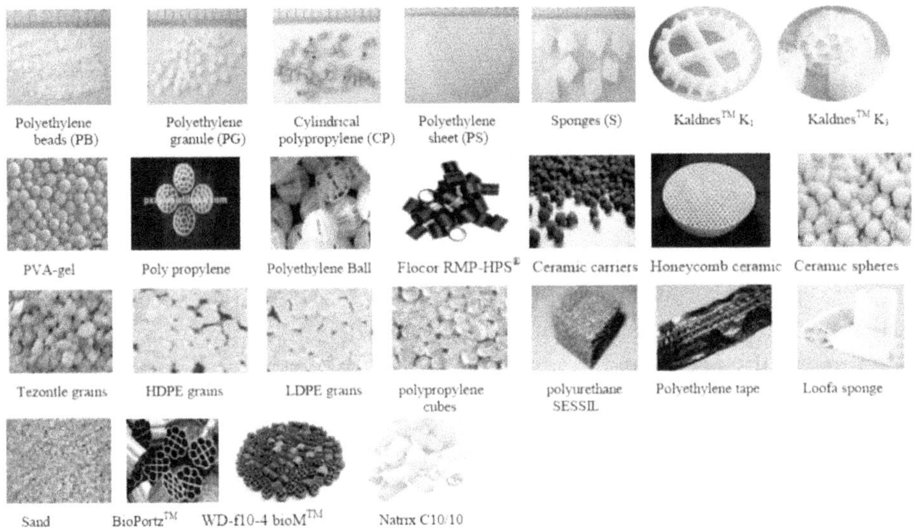

The peak flow is 2 times the average flow. The peak flow is calculated as given below:

Peak Flow

Average flow * peaking flow = 62.5 * 2 = 125 m^3/hr = 0.347 m^3/sec

Suppose operating hours is 20 hrs. The design flow is computed by the formula given below:

Design Flow

$$\frac{Flow\left(kld\right)}{operating\,hours} = \frac{1500}{20} = 75 \text{ m}^3\text{/hr}$$

Bar Screen Chamber

The retention time is 2-3 minutes of peak flow.

The volume of the screen bar is given below:

Volume (m^3) = retention time (hrs) * pcak flow (m^3/hr)

V = 0.05 * 125 = 6.25 m^3

Assume height to be 1.5 m.

$$\text{Area } (m^2) = \frac{volume}{height} = \frac{6.25}{1.5} = 4.167 \text{ m}^2 \text{ say } 4.17 \text{ m}^2$$

The Chamber dimensions can be computed as 2.1 * 2.1 * 1.5 m
Screen is always placed at an angle of 60^0 inside the chamber.

$$\text{Sin } 60^0 = \frac{perpendicular\left(chamber\ height\right)}{Hypotneous\left(screen\ height\right)}$$

Assume the screen height to be 1.15 m.

$$\text{Cross section area of the screen} = \frac{peak\ flow}{no.of\ screen * max.velocity\ through\ screen}$$

$$= \frac{0.0347}{1 * 0.78} = 0.0445 \text{ m}^2$$

Provide coarse screen of 10 mm bars with 50 mm spacing.

Oil and Grease Trap

Retention time is 5 minutes and 1 hour of peak flow for oil < 10 mg/L and upto 50 mg/L respectively.

Baffles are made up of either reinforced concrete cement of 150 mm or mild steel.

The volume of the oil and grease tank. Volume (m^3) = retention time (hrs) * peak flow (m^3/hr)

= 0.08 * 125 = 10 m^3

Assume Baffle width = 0.04 m

Baffle height = tank height – 0.3 = 3 - 0.3 = 2.7 m

Dimensions without baffle = 2 * 0.8 * 3 m

Equilization Tank

Retention time is 2-5 hours of peak flow for equalization tank.

Volume (m^3) = retention time (hrs) * peak flow (m^3/hr)

= 3.5 * 125 = 437.5 m^3

Assume tank height = 3 m
The area of tank can be calculated by the formula given below:

$$\text{Tank area} = \frac{volume}{height} = \frac{437.5}{3} = 145.83\ \text{m}^2$$

The dimensions of equilization tank = 15 * 9.7 * 3 m

Moving Bed Biofilm Reactor

The BOD load in the reactor can be given as:

$$\text{BOD load} = \frac{flow\left(kld\right) * BOD\left(ppm\right)}{1000} = \frac{1500\ {*}250}{1000} = 375\ \text{kg BOD/day}$$

BOD Reduction

Let the BOD is removed in two separate MBBR chambers.

In MBBR I = 75% of inlet BOD load = 75% of 375 kg BOD/day = 281.25 kg BOD/day

BOD load remaining in MBBR = Inlet BOD – BOD reduction

=375 – 281.25 = 93.75 kg BOD/day

BOD remaining in MBBR = 93.75 kg BOD/day

93.75 kg BOD/day will be inlet BOD load for MBBR II

In MBBR II = 85% of inlet BOD load = 85% of 93.75 = 79.96 kg BOD/day

BOD load remaining in MBBR II = inlet BOD – BOD reduction

BOD load remaining in MBBR = 93.75 – 79.96 = 13.79 kg BOD/day

Retention time is 4.5 – 6.5 hours of design flow.

The volume can be determined by the formula as discussed below:

Volume (m^3) = retention time (hrs) * design flow (m^3/hr) = 5 * 75 = 375 m^3

The media must take 30-50 percent of the volume of the tank as the microorganisms shows better efficiency. As more and more bacteria can grow on the surface of the media and microorganisms plays a major role to treat the wastewater.

The Media volume can be taken as = 30% of tank volume = 113 m^3

Total tank volume = 375 + 113 = 488 m^3

Assume Tank height can be taken as = 4 m

Tank area = 488/4 = 122 m^2

Dimensions: 11 * 11 * 4 m

Clear Water Tank

Retention time is 1.5-2 hours of design flow.

For MBBR, tank volume = retention time * design flow = 1.5 * 75 = 112.5 m^3

Assume depth of the tank is 4 m. Area = 112.5/4 = 28.13 m^2

The dimensions of the clear water tank are 5 * 5.6 * 4 m

Nitrification

Overview of Biological Nitrogen Removal Processes

Nitrification is the term used to describe the two-step biological process in which ammonia ($NH_4 - N$) is oxidised to nitrite ($NO_2 - N$) and nitrite is oxidised to nitrate ($NO_3 - N$). The effect of ammonia on receiving water with respect to DO concentration is fish toxicity. The effect of ammonia on receiving water with respect to DO concentration is fish toxicity. The need to provide nitrogen removal to prevent eutrophication or for ground water recharge or several other applications. The acceptable limit of nitrate is 45 mg/L and for nitrite is 10 mg/L in drinking water and TKN (Total Kjeldahl Nitrogen) range in municipal wastewater is 25-45 mg/L.

Process Description

Nitrification can be accomplished in both suspended growth and attached growth biological processes. For suspended growth processes, a more common approach is to achieve nitrification along with BOD removal in the same single-sludge process, consisting of an aeration tank, clarifier, and sludge recycle system. In cases where there is a significant potential for toxic and inhibitory substances in the wastewater, a two-sludge suspended system may be considered. The two-sludge system consists of two aeration tanks and two clarifiers in series with the first aeration tank/clarifier unit operated at a short SRT for BOD removal. The BOD and toxic substances are removed in the first unit, so that nitrification can proceed unhindered in the second.

In attached growth systems used for nitrification, most of the BOD must be removed before nitrifying organisms can be formed. The heterotrophic bacteria have a higher biomass yield and thus can dominate the surface area of fixed-film systems over nitrifying bacteria. Nitrification is accomplished in an attached growth reactor after BOD removal or in a separate attached growth system designed specifically for nitrification.

Microbiology

Aerobic autotrophic bacteria are responsible for nitrification in activated sludge process and biofilm processes. Nitrification is a two step process involving two groups of autotrophic bacteria. In the first stage, ammonia is oxidised to nitrite by Nitrosomonas. In the second stage, nitrite is oxidised to nitrate by Nitrobacter respectively.

Stoichiometry of Biological Nitrification

The energy yielding two-step oxidation of oxidation of ammonia to nitrate is as follows: Nitroso- bacteria:

$$2NH_4^+ + 3O_2 \rightarrow 2NO_2^- + 4H^+ + 2H_2O \quad (1)$$

Nitro-Bacteria:

$$2NO_2^- + O_2 \rightarrow 2NO_3^- \quad (2)$$

Total Oxidation Reaction:

$$NH_4^+ + 2O_2 \rightarrow NO_3^- + 2H^+ + H_2O \quad (3)$$

Based on the above total oxidation reaction, the oxygen required for complete oxidation of ammonia is 4.57 g O_2/g N oxidised with 3.43 g O_2/g used for nitrite production and 1.14 g O_2/g NO_2 oxidized.

When synthesis is considered, the amount oxygen required is less than 4.57 g O_2/g N. In addition to oxidation, oxygen is obtained from fixation of carbon dioxide and nitrogen into cell mass.

Neglecting cell tissue, the amount of alkalinity required to carry out reaction can be estimated by writing equation as follows:

$$NH_4^+ + 2HCO_3^- + 2O_2 \rightarrow NO_3^- + 2CO_2 + 3H_2O \quad (4)$$

In the above equation, for each g of ammonia nitrogen (as N) converted, 7.14 g of alkalinity as $CaCO_3$ will be required [2 x (50 g $CaCo_3$/ eq.)/14].

Along with obtaining energy, a portion of the ammonia ion is assimilated into cell tissue. The biomass synthesis reaction can be represented as follows:

$$4CO_2 + HCO_3^- + NH_4^+ + H_2O \rightarrow C_5H_7O_2N + 5O_2 \quad (5)$$

As noted obviously, the chemical formula $C_5H_7O_2N$ is used to represent the synthesized bacterial cells.

The half reactions can be used to create an equation of the overall nitrification reaction. Half reactions for cell Synthesis, Oxidation of ammonia to nitrate, and reduction of oxygen to water can be combined to create equation ($f_{s=}$0.05). Due to rounding of the coefficients, the equation dose not blance exactly; however, the error introduced by rounding is negligible (Crites and Tchobanoglous, 1998).

$$NH_4^+ + 1.863O_2 + 0.098CO_2 \rightarrow 0.0196C_5H_7NO_2 + 0.98NO_3^- + 0.0941H_2O + 1.98H^+$$

From the above equation it will be noted that for each g of ammonia nitrogen (as N) converted, 4.25 g of O_2 are utilized, 0.16 g of new cells are formed, 7.07 g of alkalinity as $CaCO_3$ are removed, and 0.08 g of inorganic carbon are utilized in the formation of new cells . The oxygen required to oxidize 1.0 g of ammonia nitrogen to nitrate (4.25) is less than the theoretical value of 4.57 g computed. Similarly, the alkalinity required for nitrification (7.07 g/g) is less than the value of 7.14 g calculated without considering the conversion of some of the ammonia to cellular nitrogen. It should be recognized that the coefficient values are depended upon the value of f_s that is used [Note that f_s = 0.05 was assumed]. Weerzernak and Gannon (1967) found that the actual total oxygen consumption was 4.33 g O_2/g N with 3.22 g O_2/g N used for ammonia oxidation and 1.11 g O_2/g N used for nitrite oxidation.

The wastewater nitrogen concentration, BOD concentration, alkalinity, temperature, and potential for toxic compound are major issues in the design of biological nitrification processes. Nitrifying bacteria need CO_2 and phosphorus for cell growth as well as trace elements. With such a low cell yield, the CO_2 in air is adequate phosphorus is seldom limiting. Trace element concentrations that have been found to stimulate the growth of nitrifying bacteria in pure culture work are: Ca = 0.50, Cu = 0.01 Mg = 0.03, Mo = 0.001, Ni = 0.10, and Zn = 1.0mg/l (poduska, 1973).

Growth Kinetics

For nitrification systems operated at temperatures at below 28⁰C ammonia-oxidation kinetics versus nitrite-oxidation kinetics are rate-limiting, so the designs are based on saturation kinetics for ammonia oxidation are given below, assuming excess DO is available.

$$\mu_m = \left(\frac{\mu_{nm} N}{K_N + N} \right) - K_{dn} \tag{6}$$

Where μ_m = specific growth rate of nitrifying bacteria, g new cells/g cells*d

μ_{nm} = maximum specific growth rate of nitrifying bacteria, g new cells/g cells*d
N = nitrogen concentration, g/m^3
K_N = half-velocity constant, substrate concentration at one half the maximum specific substrate utilization rate, g/m^3
K_{dn} = endogeneous decay coefficient for nitrifying organisms, g VSS/g VSS *d

A wide range of maximum specific growth is a function of temperature (Randall et al., 1992). At 20⁰C, μ_{nm} vary from 0.25 to 0.77 g VSS/g VSS. The μ_{nm} values for nitrifying organisms are much lower than the corresponding values for heterotrophic organisms, requiring much longer SRT values for nitrifying activated sludge systems.

SRT values for design purpose may range from 10 to 20 d at 10⁰C to 4 to 7 d at 20⁰C. Above 28⁰C, both ammonia and nitrite oxidation kinetics should be considered. At the elevated temperatures, the relative kinetics of $NH_4 - N$

and $NO_2 - N$ oxidation change, and $NO_2 - N$ will accumulate at lower SRT values.

For fully acclimated complete-mix activated sludge nitrification systems, at temperatures below 25^{0}C with sufficient DO present, the $NO_2 - N$ concentrations may be less than 0.10 mg/L as compared to $NH_4 - N$ concentrations in the range of 0.50 to 1.0 mg/L. However during the initiation of nitrification, $NO_2 - N$ will be greater than $NH_4 - N$ concentrations, as the growth of nitrite-oxidising bacteria cannot occur until the ammonia–oxidising bacteria generate nitrite.

Nitrification rates are affected by the liquids DO concentration in activated sludge. It has been observed for aerobic heterotrophic bacteria degradation of organic compounds, nitrification rates increase up to DO concentrations of 3-4 mg/L. The expression for the specific growth rate is modified as follows:

$$\mu_m = \left(\frac{\mu_{nm} N}{K_N + N} \right) \left(\frac{DO}{K_0 + DO} \right) - K_{dn} \tag{7}$$

Where DO = dissolved oxygen concentration, g/m^3

K_0 = half saturation coefficient for DO, g/m^3

Other terms are same as defined previously.

At low DO concentrations ($< 0.50 mg / L$) where nitrification rates are generally inhibited, the low DO inhibition effect has been shown to be greater for Nitrobacter than for Nitrosomonas. In such cases, incomplete nitrification will occur with increased $NO_2 - N$ concentrations in the effluent. The presence of nitrite in the effluent is particularly troublesome for plants the use chlorination for disinfection, as nitrite is readily oxidised by chlorine requiring 4 g chlorine/g $NO_2 - N$.

Environmental Factors

Nitrification is affected by a number of environmental factors including pH, toxicity, metals, and un-ionised ammonia.

Hydrogen-Ion Concentration (pH)

Nitrification is pH sensitive and rates decline significantly at pH values below 6.8. At pH values near 5.8 to 6.0, the rates may be 10 to 20 percent of the rate at pH 7.0. Optimal nitrification rates occur at pH values in the 7.5 to 8.0 range. A pH of 7.0 to 7.2 is normally used to maintain reasonable nitrification rates, and for locations with low-alkalinity waters. Alkalinity may be added in the form of lime, soda ash, sodium bicarbonate, or magnesium hydroxide depending on costs and chemical handling issues.

Toxicity

In many cases, nitrification rates are inhibited even though bacteria continue to grow and oxidise ammonia and nitrite, but at significantly reduced rates. In some cases, toxicity may be sufficient to kill the nitrifying bacteria.

Compounds that are toxic include solvent organic chemicals, amines, proteins, tannins, phenolic compounds and many more compounds. Because of the numerous compounds that can inhibit nitrification, it is difficult to find the original source of nitrification.

Metals

Metals are also the concerns for nitrifiers, and Skinner and Walker (1961) have shown complete inhibition of ammonia oxidation at 0.25 mg/L nickel, 0.25 mg/L chromium, and 0.10 mg/L copper.

Un-Ionised Ammonia

Nitrification is also inhibited by un-ionised ammonia (NH_3) or free ammonia, and un-ionised nitrous acid (HNO_2). At 20⁰C and pH 7.0, the $NH_4 - N$ concentrations at 100 mg/L and 20 mg/L may initiate inhibition of $NH_4 - N$ and $NO_2 - N$ oxidation and $NO_2 - N$ concentrations at 280 mg/L may initiate inhibition of $NO_2 - N$ oxidation.

Figure 4. Flow chart of nitrification and denitrification

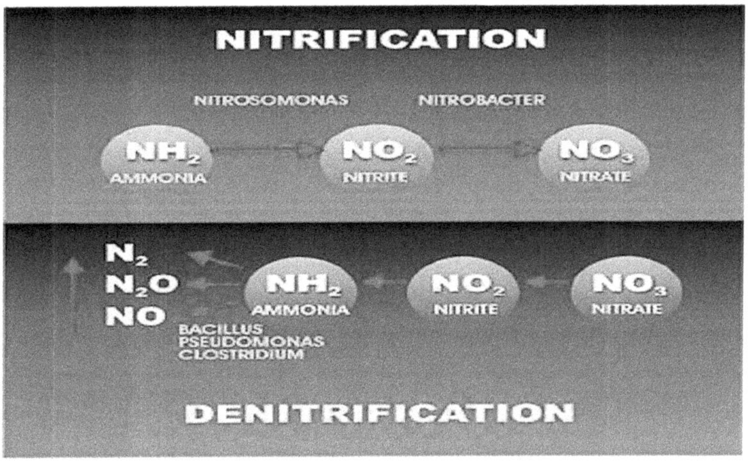

Biological Denitrification

The biological reduction of nitrate to nitric oxide, nitrous oxide, and nitrogen gas is termed as denitrification. Biological nitrogen removal is generally more cost effective and used more often. Biological nitrogen removal is used in wastewater treatment where there are concerns of eutrophication and where groundwater must be protected against elevated $NO_3 - N$ concentrationswhere wastewater treatment plant effluent is used for groundwater recharge.

Process Description

Two modes of nitrate removal can occur in biological processes, and these are termed assimilating and dissimilating nitrate reduction. Assimilating nitrate reduction involves the reduction of nitrate to ammonia for use in cell synthesis. Assimilating occur when is not available and is independent of DO concentration.

Stoichiometry of Biological Denitrification

Biological denitrification involves the biological oxidation of many organic substrates in wastewater treatment using nitrate or nitrite as the electron acceptor instead of oxygen. The nitrate reduction reactions involve the following reduction steps from nitrate to nitrite, to nitric oxide, to nitrous oxide, and to nitrogen gas:

$$NO_3^- \rightarrow NO_2^- \rightarrow NO \rightarrow N_2O \rightarrow N_2$$

In biological nitrogen removal processes, the source of electron donor:

1. Biodegradable soluble COD in influent wastewater
2. Biodegradable soluble COD produced during endogenous decay
3. Exogenous source such as methanol or acetate

The term $C_{10}H_{19}O_3N$ is used to represent the biodegradable organic matter in wastewater.

Wastewater:

$$C_{10}H_{19}O_3N + 10NO_3^- \rightarrow 5N_2 + 10CO_2 + 3H_2O + NH_3 + 10OH^- \quad (8)$$

Methanol:

$$5CH_3OH + 6NO_3^- \rightarrow 3N_2 + 5CO_2 + 7H_2O + 6OH^- \quad (9)$$

Acetate:

$$5CH_3COOH + 8NO_3^- \rightarrow 4N_2 + 10CO_2 + 6H_2O + 8OH^- \quad (10)$$

In the above heterotrophic denitrification reactions, one equivalent of alkalinity is produced per equivalent of $NO_3 - N$ reduced, which equates to 3.57 g of alkalinity (as $CaCO_3$) production per g of nitrate nitrogen reduced.

From oxidation reduction half reactions, the oxygen equivalent of using nitrate or nitrite as electron acceptors can be determined. The half reactions per mole e^- are shown as follows: For oxygen:

$$0.25O_2 + H^+ + e^- \rightarrow 0.5H_2O \quad (11)$$

For nitrate:

$$0.20NO_3^- + 1.2H^+ + e^- \rightarrow 0.1N_2 + 0.6H_2O \tag{12}$$

For nitrite:

$$0.33NO_2^- + 1.33H^+ + e^- \rightarrow 0.67H_2O + 0.17N_2 \tag{13}$$

On comparing the above half reactions for oxygen and nitrate, it should be noted that 0.25 mole of oxygen is equivalent to 0.2 mole of nitrate for electron transfer in oxidation reduction. Thus, the oxygen equivalent of nitrate is (0.25 × 32 g O_2 /mole) divided by nitrate gram equivalent (0.20 × 14 g N/mole) equals 2.86 g O_2 / g $NO_3 - N$ and correspondingly, for nitrite as the electron acceptor, the oxygen equivalent of nitrite is 1.71 g O_2 /g $NO_2 - N$.

An important design parameter for denitrification processes is the amount of bsCOD or BOD needed to provide a sufficient amount of electron donor for nitrate removal. Barth et al., 1968 estimated that 4 g of BOD is needed per g of NO_3 reduced. However, the actual value will depend on the system operating conditions and type of electron donor used for denitrification.

The amount of oxygen used per unit of bsCOD was related to the biomass yield, the bsCOD/NO_3-N ratio is similarly related to the system biomass yield. It should be noted that because of the long SRT values involving in the nitrification, the following analysis also applies for bCOD, which includes colloidal and particulate components. From a steady-state COD balance it can be shown that the bsCOD removed is oxidised or accounted for in cell growth.

$$bsCOD_u = bsCOD_{syn} + bsCOD_O$$

Where $bsCOD_u$ = bsCOD utilized, g bsCOD/d

$bsCOD_{syn}$ = bsCOD incorporated into cell synthesis, g bsCOD/d
$bsCOD_o$ = bsCOD oxidized, g bsCOD/d

For cell synthesis, the $bsCOD_{syn}$ is calculated from the net biomass yield and the ratio of $1.42 g\,O_2 / g\,VSS$. The oxygen requirement of the biomass is equal to the $bsCOD$ incorporated into biomass.

$$bsCOD_{syn} = 1.42\,Y_n\,bsCOD_u \tag{14}$$

Where Y_n = net biomass yield, g VSS/g $bsCOD_u$

$$Y_n = \frac{Y}{1+\left(K_{dn}\right)SRT} \tag{15}$$

Thus

$$bsCOD_u = bsCOD_o + 1.42\ Y_n\ bsCOD_u \tag{16}$$

Rearranging the equation

$$bsCOD_o = \left(1-1.42\ Y_n\right)bsCOD_u$$

In the above equation, $bsCOD_o$ is the COD oxidised and is equal to the oxygen equivalent of the NO_3-N used for $bsCOD$ oxidation. Hence,

$$bsCOD_o = 2.86\ NO_x$$

Where 2.86 = O_2 equivalent of NO_3-N, $g\ O_2 /\ g\ NO_3-N$

$NO_x = NO_3 - N\,\text{reduced, g / d}$

Substituting the equations

$$2.86\ NO_3 = \left(1-1.42\ Y_n\right)bsCOD_u$$

or

$$\frac{bsCOD}{NO_3-N} = \frac{2.86}{\left(1-1.42\ Y_n\right)} \tag{17}$$

$$\text{Thus, g bsCOD/g } NO_3 - N = \frac{2.86}{1 - 1.42\ Y_n}$$

Effect of Dissolved Oxygen Concentrations

Dissolved oxygen can inhibit nitrate reduction by representing the nitrate reduction enzyme. Denitrification can proceed in the presence of low bulk liquid DO concentrations.

The effect of nitrate and DO concentration on the bio-kinetics is accounted for by two correction factors expressed in the form of two saturation terms as follows:

$$r_{su} = \left(\frac{kXS}{K_s + S}\right)\left(\frac{NO_3}{K_{S,NO_3} + NO_3}\right)\left(\frac{K_0'}{K_0' + DO}\right)(\eta) \tag{18}$$

where K_0' = DO inhibition coefficient for nitrate reduction, mg/L

K_{S,NO_3} = half velocity coefficient for nitrate limited reaction, mg/L

The value of K_0' in the range from 0.1 to 0.2 mg/L have been proposed (Barker and Dold,1997).

Environmental Factors

Alkalinity is produced in denitrification reactions and the pH is generally elevated, instead of being depressed as in nitrification reactions. No significant effect on the denitrification rate of pH between 7.0 and 8.0, while Dawson and Murphy (1972) showed a decrease in the denitrification rate as the pH was decreased from 7.0 to 6.0.

Carbon Adsorption

Adsorption is the adhesion of atoms, ions or molecules from a gas, liquid or dissolved solid to a surface. This process creates a film of the adsorbate on the surface of the adsorbent.

Adsorption is present in many natural, physical, biological and chemical systems and is widely used in industrial applications such as heterogeneous catalysts, activated charcoal, capturing and using waste heat to provide cold water for air conditioning and other process requirements (adsorption chillers), synthetic resins, increasing storage capacity of carbide-derived carbons and water purification.

Development of Adsorption Isotherm

The quantity of adsorbate that can be taken up by an adsorbent is a function of both characterisitics and concentrations of adsorbate and the tempertaure. The important characteristics of adsorbate include: solubility, molecular structure, molecular weight, polarity and hydrocarbon saturation. Adsorption isotherms are developed by exposing a given amount of absorbate in a fixed volume of liquid to varying amount of activated carbon. If granular activated carbon is used, it is usually powdered to minimize adsorption times. The adsorption phase concentration is used to develop adsorption isotherms as mentioned below:

$$q_e = \frac{\left(C_o - C_e\right)V}{m} \quad (19)$$

Where q_e = adsorbent phase concentration after equilibrium, mg adsorbate/g adsorbent

C_o = Initial concentration of adsorbate, mg/L
C_e = Final equilibrium concentration of adsorbate after absorption has occurred, mg/L
V = Volume of liquid in the reactor, L
m = Mass of adsorbent,g

Freundlich Isotherm

Freundlich Isotherm is used most commonly to describe the adsorption characteristics of the activated carbon used in water and wastewater treatment. The Freundlich Isotherm is defined as follows:

$$\frac{x}{m} = K_f C_e^{1/n} \tag{20}$$

Where x /m = mass of adsorbate adsorbed per unit mass of adsorbent, mg adsorbate/ g activated carbon

K_f = Freundlich capacity factor, (mg adsorbate/ g activated carbon)(L water/ mg adsorbate)$^{1/n}$
C_e = equilibrium concentration of adsorbatein solution after adsorption, mg/L
1/n = Freundlich intensity parameter

The constants in the Freundlich isotherm can be determined by plotting log (x / m) versus log C_e and can be rewritten as:

$$\log\left(\frac{x}{m}\right) = \log K_f + \frac{1}{n}\log C_e \tag{21}$$

Langmuir Isotherm

The model applies to gases adsorbed on solid surfaces. It is a semi-empirical isotherm with a kinetic basis and was derived based on statistical thermodynamics. It is the most common isotherm equation to use due to its simplicity and its ability to fit a variety of adsorption data. It is based on four assumptions:

1. All of the adsorption sites are equivalent, and each site can only accommodate one molecule.
2. The surface is energetically homogeneous, and adsorbed molecules do not interact.
3. There are no phase transitions.
4. At the maximum adsorption, only a monolayer is formed. Adsorption only occurs on localized sites on the surface, not with other adsorbates.

The Langmuir Adsorption Isotherm is defined as:

$$\frac{x}{m} = \frac{abC_e}{1 + bC_e} \tag{22}$$

Where x / m = mass of adsorbate adsorbed per unit mass of adsorbent, mg adsorbate/ g activated carbon

a, b = empirical constants

C_e = equilibrium concentration of adsorbate in solution after adsorption, mg/L

ADSORBENTS

Characteristics and General Requirements

Activated carbon is used as an adsorbent. Adsorbents are used usually in the form of spherical pellets, rods, moldings, or monoliths with a hydrodynamic radius between 0.25 and 5 mm. They must have high abrasion resistance, high thermal stability and small pore diameters, which results in higher exposed surface area and hence high capacity for adsorption. The adsorbents must also have a distinct pore structure that enables fast transport of the gaseous vapours.

Most widely used adsorbents are listed below:

1. Silica Gel
2. Zeolites
3. Activated carbon

Most industrial adsorbents fall into one of three classes:

- **Oxygen-Containing Compounds:** Are typically hydrophilic and polar, including materials such as silica gel and zeolites.
- **Carbon-Based Compounds:** Are typically hydrophobic and non-polar, including materials such as activated carbon and graphite.
- **Polymer-Based Compounds:** Are polar or non-polar functional groups in a porous polymer matrix.

Figure 5. Activated carbon used as adsorbent

Silica Gel

Silica gel is a chemically inert, nontoxic, polar and dimensionally stable (< 400 °C or 750 °F) amorphous form of SiO_2. It is prepared by the reaction between sodium silicate and acetic acid, which is followed by a series of after-treatment processes such as aging, pickling, etc. These after-treatment methods results in various pore size distributions. Silica is used for drying of process air (e.g. oxygen, natural gas) and adsorption of heavy (polar) hydrocarbons from natural gas.

Zeolites

Zeolites are natural or synthetic crystalline aluminosilicates, which have a repeating pore network and release water at high temperature. Zeolites are polar in nature. They are manufactured by hydrothermal synthesis of sodium aluminosilicate or another silica source in an autoclave followed by ion exchange with certain cations (Na^+, Li^+, Ca^{2+}, K^+, NH_4^+). The channel diameter of zeolite cages usually ranges from 2 to 9 Å. The ion exchange process is followed by drying of the crystals, which can be pelletized with a binder to form macroporous pellets. Zeolites are applied in drying of process air, CO_2 removal from natural gas, CO removal from reforming gas, air separation, catalytic cracking, and catalytic synthesis and reforming.

Non-polar (siliceous) zeolites are synthesized from aluminum-free silica sources or by dealumination of aluminum-containing zeolites. The dealumination process is done by treating the zeolite with steam at elevated temperatures, typically greater than 500 °C (930 °F). This high temperature heat treatment breaks the aluminum-oxygen bonds and the aluminum atom is expelled from the zeolite framework.

Activated Carbon

Activated carbon is a highly porous, amorphous solid consisting of microcrystallites with a graphite lattice, usually prepared in small pellets or a powder. It is non-polar and cheap. One of its main drawbacks is that it reacts with oxygen at moderate temperatures (over 300 °C).

Activated carbon can be manufactured from carbonaceous material, including coal (bituminous, sub bituminous, and lignite), peat, wood, or nutshells (e.g., coconut). The manufacturing process consists of two phases, carbonization and activation. The carbonization process includes drying and then heating to separate by-products, including tars and other hydrocarbons from the raw material, as well as to drive off any gases generated. The process is completed by heating the material over 400 °C (750 °F) in an oxygen-free atmosphere that cannot support combustion. The carbonized particles are then "activated" by exposing them to an oxidizing agent, usually steam or carbon dioxide at high temperature. This agent burns off the pore blocking structures created during the carbonization phase and so, they develop a porous, three-dimensional graphite lattice structure.

The size of the pores developed during activation is a function of the time that they spend in this stage. Longer exposure times result in larger pore sizes. The most popular aqueous phase carbons are bituminous based because of their hardness, abrasion resistance, pore size distribution, and low cost, but their effectiveness needs to be tested in each application to determine the optimal product.

Activated carbon is used for adsorption of organic substances and non-polar adsorbates and it is also usually used for waste gas (and waste water) treatment. It is the most widely used adsorbent since most of its chemical (e.g. surface groups) and physical properties (e.g. pore size distribution and surface area) can be tuned according to what is needed. Its usefulness also derives from its large micropore (and sometimes mesopore) volume and the resulting high surface area.

Biological Phosphorous Removal

The removal of phosphorous by biological means is known as biological phosphorous removal. Phosphorous removal is done to control eutrophication because phosphorous is limiting nutrient in freshwater.

Process Occurring in the Anaerobic Zone

- Acetate is produced by fermentation of bsCOD which is dissolved degradable organic material that can be assimilated easily by the biomass. Depending on the value of for the anaerobic zone, some colloidal and particulate COD is also hydrolyzed and converted to acetate, but the amount is generally small compared to that from the bsCOD conversion.
- Using energy available from stored polyphosphates, the PAOs assimilate acetate and produce intracellular polyhydroxybutyrate (PHB) storage products. Some glycogen contained in the cell is also used. Concurrent with the acetate uptake is the release of orthophosphate as well as magnesium, potassium and calcium cations.
- The PHB content in the PAOs increase while the polyphosphate decrease.

Process Occurring in the Aerobic/Anoxic Zone

- Stored PHB is metabolized, providing energy from oxidation and carbon form new cell growth.
- Some glycogen is produced from PHB metabolism.

Figure 6. Pao phosphorous release in anaerobic zone

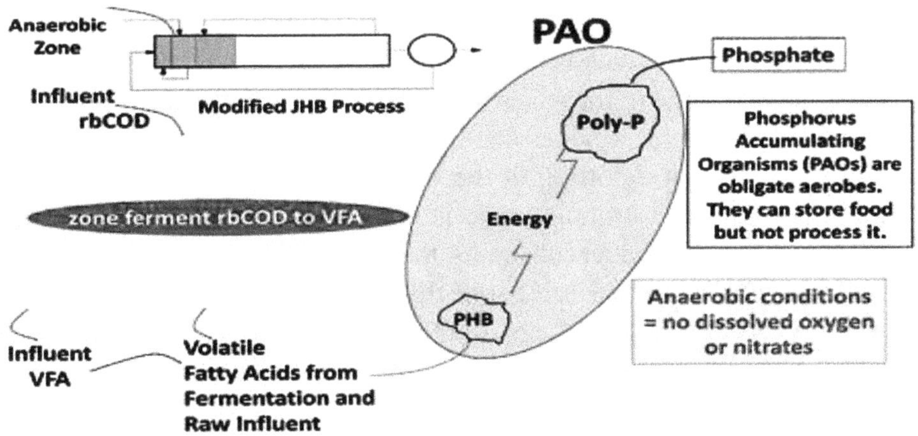

- The energy released from PHB oxidation is used to form polyphosphate bonds in cell storage so that soluble orthophosphate is removed from the solution and incorporated into polyphosphates within the bacterial cell. Cell growth also occurs due to PHB utilization and the new biomass with high polyphosphate storage accounts for phosphorous removal.
- As a portion of the biomass is wasted, stored phosphorous is removed from the biotreatment reactor for ultimate disposal with the waste sludge.

Microbiology

Phosphorous is important in cellular energy transfer mechanism via adenosine triphosphate (ATP) and polyphosphates. As energy is produced in oxidation reduction reactions, adenosine diphosphate (ADP) is converted to ATP with 7.4 kcal/mole of energy captured in the phosphate bond. As the cell uses energy, ATP is converted to ADP with phosphorous release. Many bacteria are able to store phosphorous in their cells in the form energy rich polyphosphates. The polyphosphates are contained in volutin granules within the cell along with Mg^{2+}, Ca^{2+}, and K^{+} cations.

In the anaerobic zone, high concentrations of O-PO_4 can be taken as an indication that phosphorous release by the bacteria has occurred in this zone. Significant amounts of poly-b-hydroxybutyrate (PHB) are found stored in bacteria cells, but the PHB concentration declines in the subsequent anoxic/aerobic zones and can be measured and quantified. The O-PO_4 is taken up from solution in the aerobic and anoxic zones, generally leading to very low remaining concentrations. Based on investigations of biological phosphorous removal, it was found that acetate was essential to forming the PHB under anaerobic conditions, which is advantage for the PAOs.

The anaerobic zone in the anaerobic/aerobic treatment process is termed a "selector", because it provides conditions that favour the proliferation of the PAOs, fact is that a portion of the influent bCOD is consumed by the PAOs instead of heterotrophic bacteria. The PAOs prefer low-molecular weight fermentation product substrates, the preferred food source would not be available without the anaerobic zone that provides the fermentation of the influent bsCOD to acetate. Because of the polyphosphate storage ability, the PAOs have energy available to assimilate the acetate in the anaerobic zone. Other aerobic heterotrophic bacteria have no such mechanism for acetate

Figure 7. Phosphate release in anaerobic and aerobic conditions

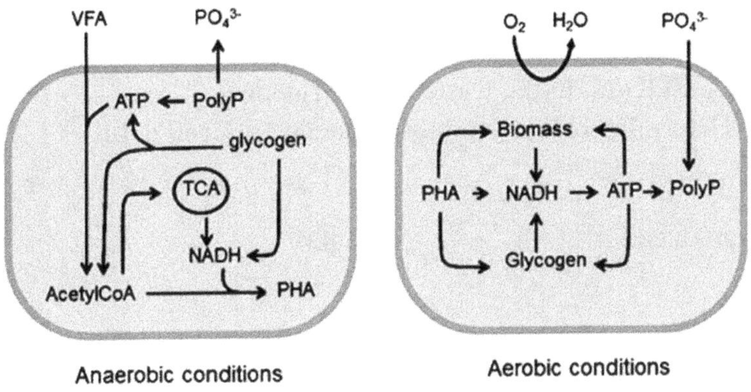

uptake, and they are starved while the PAOs assimilate COD in the anaerobic zone. The PAOs form very dense, good settling flocs in the activated sludge.

Example: Estimating the amount of phosphorous removal.

Given the following influent wastewater characteristics and the corresponding biological process information, estimate the effluent phosphorous concentrations.

1. Heterotrophic synthesis yield, Y = 0.40 g VSS/g COD
2. Endogenous decay coefficient, k_d =0.08 g VSS/g VSS*d
3. SRT = 5 days
4. Phosphorous content of PAOs = 0.30 g P/g VSS
5. Clarifier effluent VSS concentration = 8 g/m^3
6. Phosphorous content of other bacteria = 0.02 g P/g VSS

Table 1.

Influent	Concentration, g/m^3
COD	400
bCOD	300
bsCOD	100
Phosphorous	10

Solution

1. Determine phosphorous removed by PAOs due to fermentation of the 50 g rbsCOD/m^3 in the wastewater influent.
   a. Determine biomass produced neglecting cell debris

$$\text{Biomass produced} = \left[\frac{Y}{1+(k_d)SRT}\right]\text{bsCOD}$$

$$= \left[\frac{(0.40\,gVSS/gCOD)}{[1+(0.08\,gVSS/gVSS.d)(5\,d)]}\right](100\,gbsCOD/m^3) = 28.6 \text{ g VSS/m}^3$$

   b. Determine the phosphorous removed

P removed = (0.30 g P/g VSS) (28.6 g VSS/m^3) = 8.58 g/m^3

2. Determine the phosphorous removed by heterotrophs from the conversion of colloidal and particulate bCOD.
   a. Determine the COD removed

COD removed = bCOD - bsCOD = 300 – 200 g/m^3 = 100 g/m^3

$$\text{Biomass produced from bpCOD} = \left[\frac{Y}{1+(k_d)SRT}\right]\text{bpCOD}$$

$$= \left[\frac{(0.40\,gVSS/gCOD)}{[1+(0.08\,gVSS/gVSS.d)(5\,d)]}\right](100\,gbpCOD/m^3) = 28.6 \text{ g VSS/m}^3$$

   b. Determine the phosphorous removed

P removed = 0.02 g P/g VSS = 0.02(28.6) = 0.572 g/m^3

3. Determine the total phosphorous removed and effluent phosphorous concentration.

Total P removed = 8.58 + 0.572 = 9.15 g/m^3

4. Estimate P content effluent VSS.

Average P content of effluent VSS

$$= \frac{\left(0.30 \text{ g P / g VSS}\right)\left(28.6 \text{ g VSS} / m^3\right) + \left(0.02 \text{ g P / g VSS}\right)\left(28.6\text{g VSS} / m^3\right)}{\left[\left(28.6 + 28.6\right) \text{ g} / m^3\right]}$$

= 0.16 g P/g VSS

Phosphorous in effluent VSS = 0.16 (8g/m^3) = 1.28 g/m^3

Total effluent P concentration = 9.15 + 1.28 = 10.43 g/m^3

Nanotechnology in Wastewater

Conventional wastewater treatment methods include various physical, chemical and biological processes. The results of such treatment can be limited because of high investment cost or, in some cases, due to poor treatment efficiency. For that reason, new approaches are continuously being developed as a means of supplementing or replacing traditional water treatment methods. There are plenty of water purifiers available in the market which use techniques such as boiling, filtration, distillation, chlorination, sedimentation and oxidation. Currently nanotechnology plays a vital role in water purification techniques. In nanotechnology, nano membranes are used with the purpose of softening the water and removal of contaminants such as physical, biological and chemical contaminants. There are variety of techniques in nanotechnology which uses nano particles for providing safe drinking water with a high level of effectiveness. Some techniques have become commercialized. For better water purification or treatment processes nanotechnology is preferred. Many different types of nanomaterials or nanoparticles are used in water treatment processes. Nanotechnology holds great promise in remediation, desalination, filtration, purification and water treatment.

The main features that make nanoparticles effective for water treatment, It requires more surface area and small volumc. Thc higher the surface area and volume, the particles become stronger, more stable and durable. The materials may change electrical, optical, physical, chemical, or biological properties at the nano level. It makes chemical and biological reactions easier.

Nanofilteration

Membrane filtration plays an important role in removing various types of contamination and enables high level of water purification. Until recently, its biggest problem was a substantial investment cost (about 70% of the total investment cost refers to membranes). As the price is lowering, the membrane wastewater treatment process becomes more and more popular in the market, mainly due to its high efficiency in the removal of solid waste materials, monovalent and divalent ions, various pathogens, etc. Nanofiltration (with reverse osmosis, RO) is a high-pressure membrane treatment process. But unlike the RO, it requires a much lower drive pressure (7 to 14 bar), and so allows lower energy consumption. Centrifugal pumps are most often used for the pressure and circulation of wastewater within the nanomembrane. The plant consists of a large number of modules, with different membrane configurations within each module. In nanofiltration, the usual length of the module varies from 0.9 to 5.5 m, and the diameter ranges from 100 to 300 mm . The modules are installed on the stand and can be arranged either horizontally or vertically (Figure 5). For vertical installation, a smaller number of connecting pipes and fittings are required, and the footprint is smaller. Nanofiltration produces water that meets highly stringent requirements in terms of water reuse. Since this process is highly efficient in the removal of organic and inorganic substances, bacteria and viruses, the need for subsequent disinfection of water is minimal.

Nanomaterials for Water Disinfection

In addition to having excellent adsorption and catalytic properties, some nanomaterials have proven to have great antimicrobial activity as well. Such materials include chitosan, silver nanoparticles, titanium dioxide, fullerene nanoparticles, carbon nanotubes, etc. All these nanomaterials are mild oxidants and are relatively inert in water, and are therefore note expected to create harmful by-products. There are several ways of applying the nanomaterials in water disinfection processes:

- Direct action on (bacterial) cells in the sense of preventing electron passage through the membrane - break through the cell membrane - oxidation of some cellular components
- Hydroxyl radicals (within the action of nanoparticles as photocatalysts)
- The formation of dissolved metal ions that can cause damage to cellular components.

However, there are some limitations regarding the use of nanotechnology in wastewater disinfection processes. For some nanomaterials to be effective in removing different types of microorganisms, they must be in direct contact with the cell membrane of bacteria, viruses, etc. Therefore, some nanomaterials (carbon nanotubes) need to be strongly connected to the reactive surface. Also, the deficiency of nanotechnology in disinfection processes is that there is no residual, i.e., subsequent antimicrobial activity in wastewater (such as in the case of chlorine use). The great advantage of conventional disinfection procedures is currently low cost. Nanotechnological processes will become more competitive after standardisation of the production of nanomaterials and following reduction of total labour costs.

Nanomaterials for Adsorption of Pollutants

Nanoparticles possess two important characteristics that make them very good adsorbents. These are the large specific surface of nanomaterials and surface multi-functionality or the ability to easily chemically react and bind to different adjacent atoms and molecules. These characteristics make nanoparticles not only effective adsorbents for various contaminants in wastewater but also allow for long-term stability, as this also results in adsorbent degradation (with the addition of catalytic properties of nanoparticles) and improves the adsorption efficiency. With the discovery of carbon nanotubes, a new carbon-based adsorption material was introduced to the world. Compared to the best known such material - activated carbon - carbon nanotubes possess approximately the same large specific surface, but their great advantage lies in the structure of nanomaterials and a much better arrangement of carbon atoms. In addition, nanomaterials possess unique mechanical, electrical, chemical, optical and many other characteristics that allow them to have much better adsorption properties for some contaminants (heavy metals and organic pollutants).

Besides carbon nanotubes, metal based nanoparticles also have adsorption characteristics. The most common metal oxides used as adsorbents are iron oxides (Fe_xO_y), silicon (Si), titanium (Ti) and tungsten (W). They are mainly used for adsorption of heavy metals and radionuclides (unstable nuclides). The adsorption process is based on the electrostatic interaction of dissolved metals in wastewater and the nanoadsorbent surface. Changing the pH of the solution can significantly affect the strength of this interaction. Thus, the surface of the nanoadsorbent may be:

- Acidic, with positive charge attracting anions
- Basic, with a negative charge attracting cations from waste water

Environmental and Human Health Effects of Nanomaterials Used in Wastewater Treatment

Contemporary nanotechnology research in the sphere of wastewater treatment should not solely be based on the potential for improving properties of nanomaterials used in wastewater treatment, their efficiency, or standardization. A very important factor that is highly significant for broader commercial use of these products in the future is their impact on human health and the ecosystem. In almost all forms of application of this technology, the most critical part is the presence of nanomaterials in wastewater, because of their potential environmental impact. Additionally, once nanoparticles come to nature, their interaction with chemical substances in the environment can often have negative consequences.

The largest quantities of nanomaterials that appear in the environment will eventually end up in soil, while smaller amounts of nanoparticles are present in water and air. One of the largest sources of nanomaterials in the soil is the sludge generated in wastewater treatment plants. After wastewater treatment, total pollution (including nanoparticles) accumulates in sludge, which is then transported and handled in various ways. A major problem in the treatment phase is that various types of polyelectrolytes are added to the wastewater for the purposes of flocculation and sedimentation of particles, and they directly affect the abovementioned properties of nanoparticles: reactivity, dispersion, mobility, etc. The sludge is then used for various purposes and it ends up in the environment where it is either deposited at landfills, used as soil improver in agriculture, etc.

REFERENCES

Adam, C., Gnirss, R., Lesjean, B., Buisson, H., & Kraume, M. (2002). Enhanced biological phosphorus removal in membrane bioreactors. *Water Science and Technology*, *46*(4-5), 281–286. doi:10.2166/wst.2002.0606 PMID:12361022

Barker, P. S., & Dold, P. L. (1997). General model for biological nutrient removal in activated sludge systems: Model presentation. *Water Environment Research*, *69*(5), 969–984. doi:10.2175/106143097X125669

Barth, E. F., Brenner, R. C., & Lewis, R. F. (1968). Chemical biological control of nitrogen and phosphorous in wastewater effluent. *Journal - Water Pollution Control Federation*, *40*, 2040.

Chang, I.S., Le Clech, P., Jefferson, B., & Judd, S. (2002). Membrane Fouling in Membrane Bioreactors for Wastewater Treatment. *Journal of Environmental Engineering*, 1019-1029.

Crites, R., & Tchobanoglous, G. (1998). *Small and Decentralized Wastewater Management Systems*. New York: McGraw-Hill.

Dawson, R. N., & Murphy, K. L. (1972). The temperature dependency of biological denitrification. *Water Research*, *6*(1), 71–83. doi:10.1016/0043-1354(72)90174-1

Ersu, C. B., Ong, S. K., Arslankaya, E., & Brown, P. (2008). Comparison of recirculation configurations for biological nutrient removal in a membrane bioreactor. *Water Research*, *42*(6–7), 1651–1663. doi:10.1016/j.watres.2007.10.022 PMID:17991508

Galil, N. I., Malachi, K. B. D., & Sheindorf, C. (2009). Biological nutrient removal in membrane biological reactors. *Environmental Engineering Science*, *26*(4), 817–824. doi:10.1089/ees.2008.0234

Jianlong, W., Hanchang, S., & Yi, Q. (2000). Wastewater treatment in a hybrid biological reactor (HBR): Effect of organic loading rates. *Process Biochemistry*, *36*(4), 297–303. doi:10.1016/S0032-9592(00)00153-9

Kermani, M., Bina, B., Movahedian, H., Amin, M. M., & Nikaein, M. (2008). Application of moving bed biofilm process for biological organics and nutrients removal from municipal wastewater. *American Journal of Environmental Sciences*, *4*(6), 675–682. doi:10.3844/ajessp.2008.675.682

Kraume, M., Bracklow, U., Vocks, M., & Drews, A. (2005). Nutrients removal in MBRs for municipal wastewater treatment. *Water Science and Technology*, *51*(6-7), 391–402. doi:10.2166/wst.2005.0661 PMID:16004001

Lesjean, B., Gnirss, R., & Adam, C. (2002). Process configurations adapted to membrane bioreactors for enhanced biological phosphorous and nitrogen removal. *Desalination*, *149*(1–3), 217–224. doi:10.1016/S0011-9164(02)00762-2

Odegaard, H. (1999). The moving bed biofilm Reactor. In Water environmental engineering and reuse of water. Hokkaido Press.

Poduska, R. A. (1973). *A Dynamic model of nitrification for the activated sludge process* (Ph.D. Thesis). Clemson University.

Rahimi, Y., Torabian, A., Mehrdadi, N., Rezaie, M. H., Pezeshk, H., & Bidhendi, G. R. N. (2011). Optimizing aeration rates for minimizing membrane fouling and its effect on sludge characteristics in a moving bed membrane bioreactor. *Journal of Hazardous Materials*, *186*(2-3), 1097–1102. doi:10.1016/j.jhazmat.2010.11.117 PMID:21168965

Randall, Barnard, & Stensel. (1992). *Design and retrofit of wastewater treatment plants for biological nutrient removal* (vol. 5). Water Quality Management Library.

Sharrer, M. J., Tal, Y., Ferrier, D., Hankins, J. A., & Summerfelt, S. T. (2007). Membrane biological reactor treatment of a saline backwash flow from a recirculating aquaculture system. *Aquacultural Engineering*, *36*(2), 159–176. doi:10.1016/j.aquaeng.2006.10.003

Shore, J. L., M'Coy, W. S., Gunsch, C. K., & Deshusses, M. A. (2012). Application of a moving bed biofilm reactor for tertiary ammonia treatment in high temperature industrial wastewater. *Bioresource Technology*, *112*, 51–60. doi:10.1016/j.biortech.2012.02.045 PMID:22444639

Skinner, F. A., & Walker, N. (1961). growth of nitrosomonas europaea in batch and continuous culture. *Archives of Microbiology*, *38*, 339.

Smith, C. V., DiGregorio, D., & Talcott, R. M. (1969). The use of ultrafiltration membranes for activated sludge separation. *Proceedings of the 24th Annual Purdue Industrial Waste Conference*.

Song, K. G., Cho, J., Ko, E. T., & Alan, K. H. (2004). *Characteristics of nitrogen and phosphorous removal in a sequencing anoxic/anaerobic membrane bioreactor (SAM) process. 4th IWA World Water Congress*, Marrakech, Morocco.

Sonune, A., & Ghate, R. (2004). Developments in wastewater treatment methods. *Desalination*, *167*, 55–63. doi:10.1016/j.desal.2004.06.113

Tavares, C. R. G., Santanna, G. L. Jr, & Capdeville, B. (1995). The effect of air superficial velocity on biofilm accumulation in a three phase fluidized-bed reactor. *Water Resources*, *29*(10), 2293–2298.

Yuan, L. M., Zhang, C. Y., Zhang, Y. Q., Ding, Y., & Xi, D. L. (2008). Biological nutrient removal using an alternating of anoxic and anaerobic membrane bioreactor (AAAM) process. *Desalination*, *221*(1–3), 566–575. doi:10.1016/j.desal.2007.01.118

Related Readings

To continue IGI Global's long-standing tradition of advancing innovation through emerging research, please find below a compiled list of recommended IGI Global book chapters and journal articles in the areas of wastewater treatment, environmental sustainability, and nanotechnology. These related readings will provide additional information and guidance to further enrich your knowledge and assist you with your own research.

Affam, A. C., & Chaudhuri, M. (2019). Comparative Study of Advance Oxidation Processes for Treatment of Pesticide Wastewater. In H. Aziz & S. Abu Amr (Eds.), *Advanced Oxidation Processes (AOPs) in Water and Wastewater Treatment* (pp. 261–323). Hershey, PA: IGI Global. doi:10.4018/978-1-5225-5766-1.ch011

Ahmedi, F., Ahmedi, L., O'Flynn, B., Kurti, A., Tahirsylaj, S., Bytyçi, E., ... Salihu, A. (2018). InWaterSense: An Intelligent Wireless Sensor Network for Monitoring Surface Water Quality to a River in Kosovo. *International Journal of Agricultural and Environmental Information Systems*, *9*(1), 39–61. doi:10.4018/IJAEIS.2018010103

Akash, P. V. M., Babu, N., & Navneet. (2018). Bioremediation of Environmental Pollutants. In V. Pathak, & Navneet (Eds.), Handbook of Research on Microbial Tools for Environmental Waste Management (pp. 80-104). Hershey, PA: IGI Global. doi:10.4018/978-1-5225-3540-9.ch005

Akbar, N. A., Matsin, Z. B., & Ramli, S. F. (2019). Groundwater Treatment via Ozonation. In H. Aziz & S. Abu Amr (Eds.), *Advanced Oxidation Processes (AOPs) in Water and Wastewater Treatment* (pp. 200–225). Hershey, PA: IGI Global. doi:10.4018/978-1-5225-5766-1.ch009

Ali, D. A., Palaniandy, P., & Feroz, S. (2019). Advanced Oxidation Processes (AOPs) to Treat the Petroleum Wastewater. In H. Aziz & S. Abu Amr (Eds.), *Advanced Oxidation Processes (AOPs) in Water and Wastewater Treatment* (pp. 99–122). Hershey, PA: IGI Global. doi:10.4018/978-1-5225-5766-1.ch005

Amorim, C. L., Moreira, I. S., Duque, A. F., van Loosdrecht, M. C., & Castro, P. M. (2017). Aerobic Granular Sludge: Treatment of Wastewaters Containing Toxic Compounds. In Á. Val del Río, J. Campos Gómez, & A. Mosquera Corral (Eds.), *Technologies for the Treatment and Recovery of Nutrients from Industrial Wastewater* (pp. 231–263). Hershey, PA: IGI Global. doi:10.4018/978-1-5225-1037-6.ch009

Anderson, S., & Rahman, P. K. (2018). Bioprocessing Requirements for Bioethanol: Sugarcane vs. Sugarcane Bagasse. In V. Pathak, & Navneet (Eds.), Handbook of Research on Microbial Tools for Environmental Waste Management (pp. 48-56). Hershey, PA: IGI Global. doi:10.4018/978-1-5225-3540-9.ch003

Ara, T., Bashir, R., Chisti, H., & Rangreez, T. A. (2019). Wastewater Pollution From the Industries. In A. Hussain & S. Ahmed (Eds.), *Advanced Treatment Techniques for Industrial Wastewater* (pp. 98–113). Hershey, PA: IGI Global. doi:10.4018/978-1-5225-5754-8.ch007

Aris, A., Jusoh, M. N., & Wahab, N. S. (2019). Applications of Advanced Oxidation Processes in Palm Oil Mill Effluent Treatment. In H. Aziz & S. Abu Amr (Eds.), *Advanced Oxidation Processes (AOPs) in Water and Wastewater Treatment* (pp. 123–149). Hershey, PA: IGI Global. doi:10.4018/978-1-5225-5766-1.ch006

Aziz, N. A., & Palaniandy, P. (2019). Photocatalysis (TiO2/Solar) in Water and Wastewater Treatment. In H. Aziz & S. Abu Amr (Eds.), *Advanced Oxidation Processes (AOPs) in Water and Wastewater Treatment* (pp. 171–199). Hershey, PA: IGI Global. doi:10.4018/978-1-5225-5766-1.ch008

Babu, N., & Pathak, V. M. Akash, & Navneet, (2018). Biosorption of Heavy Metals: Biological Approach to Control the Industrial Waste. In V. Pathak, & Navneet (Eds.), Handbook of Research on Microbial Tools for Environmental Waste Management (pp. 270-281). Hershey, PA: IGI Global. doi:10.4018/978-1-5225-3540-9.ch013

Bagheri, S., & Julkapli, N. M. (2017). Biomass-Derived Activated Carbon: Synthesis, Functionalized, and Photocatalysis Application. In T. Saleh (Ed.), *Advanced Nanomaterials for Water Engineering, Treatment, and Hydraulics* (pp. 162–199). Hershey, PA: IGI Global. doi:10.4018/978-1-5225-2136-5.ch007

Banerjee, P., Mukhopadhayay, A., & Das, P. (2016). Advances in Bioremediation for Removal of Toxic Dye from Different Streams of Wastewater. In A. Rathoure & V. Dhatwalia (Eds.), *Toxicity and Waste Management Using Bioremediation* (pp. 266–278). Hershey, PA: IGI Global. doi:10.4018/978-1-4666-9734-8.ch013

Basak, B., & Dey, A. (2016). Bioremediation Approaches for Recalcitrant Pollutants: Potentiality, Successes and Limitation. In A. Rathoure & V. Dhatwalia (Eds.), *Toxicity and Waste Management Using Bioremediation* (pp. 178–197). Hershey, PA: IGI Global. doi:10.4018/978-1-4666-9734-8.ch009

Basheer, T., & Umesh, M. (2018). Valorization of Tannery Solid Waste Materials Using Microbial Techniques: Microbes in Tannery Solid Waste Management. In V. Pathak, & Navneet (Eds.), Handbook of Research on Microbial Tools for Environmental Waste Management (pp. 127-145). Hershey, PA: IGI Global. doi:10.4018/978-1-5225-3540-9.ch007

Becker, J. A., & Stefanakis, A. I. (2016). Pharmaceuticals and Personal Care Products as Emerging Water Contaminants. In A. McKeown & G. Bugyi (Eds.), *Impact of Water Pollution on Human Health and Environmental Sustainability* (pp. 81–100). Hershey, PA: IGI Global. doi:10.4018/978-1-4666-9559-7.ch004

Belmonte, M., Fajardo, C., Toledo-Alarcón, J. B., Heredia, D. V., Jorquera, L., Méndez, R., ... Ruiz-Filippi, G. (2017). Autotrophic Denitrification Processes. In Á. Val del Río, J. Campos Gómez, & A. Mosquera Corral (Eds.), *Technologies for the Treatment and Recovery of Nutrients from Industrial Wastewater* (pp. 147–173). Hershey, PA: IGI Global. doi:10.4018/978-1-5225-1037-6.ch006

Bender, S. R. (2016). Floodplain Infrastructure and the Toxic Tide. In A. McKeown & G. Bugyi (Eds.), *Impact of Water Pollution on Human Health and Environmental Sustainability* (pp. 150–173). Hershey, PA: IGI Global. doi:10.4018/978-1-4666-9559-7.ch007

Benjamin, S. R., de Lima, F., & Rathoure, A. K. (2016). Genetically Engineered Microorganisms for Bioremediation Processes: GEMs for Bioremediaton. In A. Rathoure & V. Dhatwalia (Eds.), *Toxicity and Waste Management Using Bioremediation* (pp. 113–140). Hershey, PA: IGI Global. doi:10.4018/978-1-4666-9734-8.ch006

Bhakta, J. N. (2016). Microbial Response against Metal Toxicity. In A. Rathoure & V. Dhatwalia (Eds.), *Toxicity and Waste Management Using Bioremediation* (pp. 75–96). Hershey, PA: IGI Global. doi:10.4018/978-1-4666-9734-8.ch004

Bhatt, P. (2018). Insilico Tools to Study the Bioremediation in Microorganisms. In V. Pathak, & Navneet (Eds.), Handbook of Research on Microbial Tools for Environmental Waste Management (pp. 389-395). Hershey, PA: IGI Global. doi:10.4018/978-1-5225-3540-9.ch018

Bogatu, C., & Leszczynska, D. (2016). Transformation of Nanomaterials in Environment: Surface Activation of SWCNTs during Disinfection of Water with Chlorine. *Journal of Nanotoxicology and Nanomedicine*, *1*(1), 45–57. doi:10.4018/JNN.2016010104

Bonmatí-Blasi, A., Cerrillo-Moreno, M., & Riau-Arenas, V. (2017). Systems Based on Physical-Chemical Processes: Nutrient Recovery for Cycle Closure. In Á. Val del Río, J. Campos Gómez, & A. Mosquera Corral (Eds.), *Technologies for the Treatment and Recovery of Nutrients from Industrial Wastewater* (pp. 43–75). Hershey, PA: IGI Global. doi:10.4018/978-1-5225-1037-6.ch003

Cema, G., & Sochacki, A. (2017). Treatment of Landfill Leachate by Anammox Process. In Á. Val del Río, J. Campos Gómez, & A. Mosquera Corral (Eds.), *Technologies for the Treatment and Recovery of Nutrients from Industrial Wastewater* (pp. 290–311). Hershey, PA: IGI Global. doi:10.4018/978-1-5225-1037-6.ch011

Cervantes, F. J., Cuervo-López, F., & Hernández, J. G. (2017). Fundamentals of the Biological Processes for Nitrogen Removal. In Á. Val del Río, J. Campos Gómez, & A. Mosquera Corral (Eds.), *Technologies for the Treatment and Recovery of Nutrients from Industrial Wastewater* (pp. 112–146). Hershey, PA: IGI Global. doi:10.4018/978-1-5225-1037-6.ch005

Chaudhry, S., & Garg, S. (2019). Industrial Wastewater Pollution and Advanced Treatment Techniques. In A. Hussain & S. Ahmed (Eds.), *Advanced Treatment Techniques for Industrial Wastewater* (pp. 74–97). Hershey, PA: IGI Global. doi:10.4018/978-1-5225-5754-8.ch006

Chaurasia, P. K., Bharati, S. L., & Mani, A. (2018). Enzymatic Treatment of Petroleum-Based Hydrocarbons. In V. Pathak, & Navneet (Eds.), Handbook of Research on Microbial Tools for Environmental Waste Management (pp. 396-408). Hershey, PA: IGI Global. doi:10.4018/978-1-5225-3540-9.ch019

Cook, C., & Gude, V. G. (2017). Characteristics of Chitosan Nanoparticles for Water and Wastewater Treatment: Chitosan for Water Treatment. In T. Saleh (Ed.), *Advanced Nanomaterials for Water Engineering, Treatment, and Hydraulics* (pp. 223–261). Hershey, PA: IGI Global. doi:10.4018/978-1-5225-2136-5.ch009

Cuker, B., & Bugyi, G. (2016). The Fundamentals of Water and Natural Waters. In A. McKeown & G. Bugyi (Eds.), *Impact of Water Pollution on Human Health and Environmental Sustainability* (pp. 1–28). Hershey, PA: IGI Global. doi:10.4018/978-1-4666-9559-7.ch001

de la Varga, D., Soto, M., Arias, C. A., van Oirschot, D., Kilian, R., Pascual, A., & Álvarez, J. A. (2017). Constructed Wetlands for Industrial Wastewater Treatment and Removal of Nutrients. In Á. Val del Río, J. Campos Gómez, & A. Mosquera Corral (Eds.), *Technologies for the Treatment and Recovery of Nutrients from Industrial Wastewater* (pp. 202–230). Hershey, PA: IGI Global. doi:10.4018/978-1-5225-1037-6.ch008

Dhatwalia, V. K., & Nanda, M. (2016). Biodegradation of Phenol: Mechanisms and Applications. In A. Rathoure & V. Dhatwalia (Eds.), *Toxicity and Waste Management Using Bioremediation* (pp. 198–214). Hershey, PA: IGI Global. doi:10.4018/978-1-4666-9734-8.ch010

Di Paola, N., Spanò, R., Caldarelli, A., & Vona, R. (2016). Fighting Ecomafias: The Role of Biotech Networks in Achieving Sustainability. In A. Rathoure & V. Dhatwalia (Eds.), *Toxicity and Waste Management Using Bioremediation* (pp. 295–311). Hershey, PA: IGI Global. doi:10.4018/978-1-4666-9734-8.ch015

Dima, J. B., & Zaritzky, N. (2017). Performance of Chitosan Micro/Nanoparticles to Remove Hexavalent Chromium From Residual Water. In T. Saleh (Ed.), *Advanced Nanomaterials for Water Engineering, Treatment, and Hydraulics* (pp. 262–288). Hershey, PA: IGI Global. doi:10.4018/978-1-5225-2136-5.ch010

Dubey, S. K., Hussain, A., & Ajnavi, S. (2019). Environmental Recycling System (ERS): An Emerging Approach to Solid Waste Management. In A. Hussain & S. Ahmed (Eds.), *Advanced Treatment Techniques for Industrial Wastewater* (pp. 225–237). Hershey, PA: IGI Global. doi:10.4018/978-1-5225-5754-8.ch013

Farraji, H. (2018). Phytoremediation of Nitrogen and Phosphorus in Municipal Wastewater by Cyperus alternifolius Planted Constructed Wetland. In V. Pathak, & Navneet (Eds.), Handbook of Research on Microbial Tools for Environmental Waste Management (pp. 146-163). Hershey, PA: IGI Global. doi:10.4018/978-1-5225-3540-9.ch008

Farraji, H., Zaman, N. Q., Vakili, M., & Faraji, H. (2016). Overpopulation and Sustainable Waste Management. *International Journal of Sustainable Economies Management*, *5*(3), 13–36. doi:10.4018/IJSEM.2016070102

Fra-Vázquez, A., & Valenzuela-Heredia, D. (2017). Biogeochemical Cycles of Nitrogen and Phosphorus: Implications of Anthropogenic Activities. In Á. Val del Río, J. Campos Gómez, & A. Mosquera Corral (Eds.), *Technologies for the Treatment and Recovery of Nutrients from Industrial Wastewater* (pp. 1–20). Hershey, PA: IGI Global. doi:10.4018/978-1-5225-1037-6.ch001

Gatt, K. (2016). Youths' Social Traits in Water Management as a Precursor for Good Water Governance. *International Journal of Information Systems and Social Change*, *7*(3), 16–26. doi:10.4018/IJISSC.2016070102

Gupta, C., & Prakash, D. (2016). Novel Bioremediation Methods in Waste Management: Novel Bioremediation Methods. In A. Rathoure & V. Dhatwalia (Eds.), *Toxicity and Waste Management Using Bioremediation* (pp. 141–157). Hershey, PA: IGI Global. doi:10.4018/978-1-4666-9734-8.ch007

Hassan, S. H., & Halim, A. A. (2019). Water Quality Legislation and Regulation. In H. Aziz & S. Abu Amr (Eds.), *Advanced Oxidation Processes (AOPs) in Water and Wastewater Treatment* (pp. 30–45). Hershey, PA: IGI Global. doi:10.4018/978-1-5225-5766-1.ch002

Hilles, A. H., Abu Amr, S. S., Aziz, H. A., & Bashir, M. J. (2019). Advanced Oxidation Processes for Water and Wastewater Treatment: An Introduction. In H. Aziz & S. Abu Amr (Eds.), *Advanced Oxidation Processes (AOPs) in Water and Wastewater Treatment* (pp. 46–69). Hershey, PA: IGI Global. doi:10.4018/978-1-5225-5766-1.ch003

Hivrale, A. U., Pawar, P. K., Rane, N. R., & Govindwar, S. P. (2016). Application of Genomics and Proteomics in Bioremediation. In A. Rathoure & V. Dhatwalia (Eds.), *Toxicity and Waste Management Using Bioremediation* (pp. 97–112). Hershey, PA: IGI Global. doi:10.4018/978-1-4666-9734-8.ch005

Hussain, A., Bhattacharya, A., & Ahmed, A. (2019). Plastic Waste Pollution and Its Management in India: A Review. In A. Hussain & S. Ahmed (Eds.), *Advanced Treatment Techniques for Industrial Wastewater* (pp. 62–73). Hershey, PA: IGI Global. doi:10.4018/978-1-5225-5754-8.ch005

Ibrahim, N., Zainal, S. F., & Aziz, H. A. (2019). Application of UV-Based Advanced Oxidation Processes in Water and Wastewater Treatment. In H. Aziz & S. Abu Amr (Eds.), *Advanced Oxidation Processes (AOPs) in Water and Wastewater Treatment* (pp. 384–414). Hershey, PA: IGI Global. doi:10.4018/978-1-5225-5766-1.ch014

Jagdale, S., & Chabukswar, A. (2016). Phyto-Remediation: Using Plants to Clean Up Soils: Phyto-Remediation. In A. Rathoure & V. Dhatwalia (Eds.), *Toxicity and Waste Management Using Bioremediation* (pp. 215–235). Hershey, PA: IGI Global. doi:10.4018/978-1-4666-9734-8.ch011

Kamaruddin, M. A. (2019). Advanced Oxidation Processes (AOPs) in Landfill Leachate Treatment. In H. Aziz & S. Abu Amr (Eds.), *Advanced Oxidation Processes (AOPs) in Water and Wastewater Treatment* (pp. 355–383). Hershey, PA: IGI Global. doi:10.4018/978-1-5225-5766-1.ch013

Kaur, A. (2017). Applications of Nanomaterials for Water Treatment: A Future Avenue. In T. Saleh (Ed.), *Advanced Nanomaterials for Water Engineering, Treatment, and Hydraulics* (pp. 289–304). Hershey, PA: IGI Global. doi:10.4018/978-1-5225-2136-5.ch011

Kehinde, F. O., Abu Amr, S. S., & Aziz, H. A. (2019). Application of Persulfate in Textile Wastewater Treatment. In H. Aziz & S. Abu Amr (Eds.), *Advanced Oxidation Processes (AOPs) in Water and Wastewater Treatment* (pp. 70–98). Hershey, PA: IGI Global. doi:10.4018/978-1-5225-5766-1.ch004

Khelfi, A. (2018). Sources of Air Pollution. In V. Pathak, & Navneet (Eds.), Handbook of Research on Microbial Tools for Environmental Waste Management (pp. 220-258). Hershey, PA: IGI Global. doi:10.4018/978-1-5225-3540-9.ch011

Khelfi, A. (2018). Sources of Soil Pollution. In V. Pathak, & Navneet (Eds.), Handbook of Research on Microbial Tools for Environmental Waste Management (pp. 190-218). Hershey, PA: IGI Global. doi:10.4018/978-1-5225-3540-9.ch010

Khelfi, A. (2019). Sources of Groundwater Pollution. In A. Hussain & S. Ahmed (Eds.), *Advanced Treatment Techniques for Industrial Wastewater* (pp. 177–210). Hershey, PA: IGI Global. doi:10.4018/978-1-5225-5754-8.ch011

Koli, S. K., & Hussain, A. (2019). Status of Electronic Waste Management in India: A Review. In A. Hussain & S. Ahmed (Eds.), *Advanced Treatment Techniques for Industrial Wastewater* (pp. 238–250). Hershey, PA: IGI Global. doi:10.4018/978-1-5225-5754-8.ch014

Kulkarni, A. G., & Vedamurthy, A. B. (2018). Cellulolytic Microorganisms: A Review. In V. Pathak, & Navneet (Eds.), Handbook of Research on Microbial Tools for Environmental Waste Management (pp. 34-47). Hershey, PA: IGI Global. doi:10.4018/978-1-5225-3540-9.ch002

Kumar, P. S., & C., F. C. (2019). Soil Bioremediation Techniques. In A. Hussain, & S. Ahmed (Eds.), *Advanced Treatment Techniques for Industrial Wastewater* (pp. 35-50). Hershey, PA: IGI Global. doi:10.4018/978-1-5225-5754-8.ch003

Kumar, P. S., & Saravanan, A. (2019). Water Pollutants and Their Removal Techniques. In A. Hussain & S. Ahmed (Eds.), *Advanced Treatment Techniques for Industrial Wastewater* (pp. 114–133). Hershey, PA: IGI Global. doi:10.4018/978-1-5225-5754-8.ch008

Kumar, P. S., & Yaashikaa, P. R. (2019). Agriculture Pollution. In A. Hussain & S. Ahmed (Eds.), *Advanced Treatment Techniques for Industrial Wastewater* (pp. 134–154). Hershey, PA: IGI Global. doi:10.4018/978-1-5225-5754-8.ch009

Laha, S. D., Dutta, K., & Kundu, P. P. (2018). Biodegradation of Low Density Polyethylene Films. In V. Pathak, & Navneet (Eds.), Handbook of Research on Microbial Tools for Environmental Waste Management (pp. 282-318). Hershey, PA: IGI Global. doi:10.4018/978-1-5225-3540-9.ch014

Manyuchi, M. M., Muzenda, E., & Mbohwa, C. (2018). Design and Development of a Sanitary Landfill for Low Income Countries for Optimal Waste Management. In V. Pathak, & Navneet (Eds.), Handbook of Research on Microbial Tools for Environmental Waste Management (pp. 373-388). Hershey, PA: IGI Global. doi:10.4018/978-1-5225-3540-9.ch017

Mbuh, M. J., Houser, P. R., & Heidari, A. (2016). Water Quality Estimation Using Combined Water Chemistry and Field Spectroscopy in the Shenandoah River, Virginia. *International Journal of Applied Geospatial Research*, *7*(2), 14–37. doi:10.4018/ijagr.2016040102

McKeown, A. E. (2016). Nurses, Healthcare, and Environmental Pollution and Solutions: Breaking the Cycle of Harm. In A. McKeown & G. Bugyi (Eds.), *Impact of Water Pollution on Human Health and Environmental Sustainability* (pp. 174–197). Hershey, PA: IGI Global. doi:10.4018/978-1-4666-9559-7.ch008

Mobin, R., Chisti, H., Rangreez, T. A., Bashir, R., & Najar, A. A. (2019). Pesticidal Pollution. In A. Hussain & S. Ahmed (Eds.), *Advanced Treatment Techniques for Industrial Wastewater* (pp. 155–176). Hershey, PA: IGI Global. doi:10.4018/978-1-5225-5754-8.ch010

Mohamed, Z. (2016). Cyanobacterial Toxins in Water Sources and Their Impacts on Human Health. In A. McKeown & G. Bugyi (Eds.), *Impact of Water Pollution on Human Health and Environmental Sustainability* (pp. 120–149). Hershey, PA: IGI Global. doi:10.4018/978-1-4666-9559-7.ch006

Mojiri, A., Ziyang, L., Hui, W., & Gholami, A. (2019). Concentrated Landfill Leachate Treatment by Electro-Ozonation. In H. Aziz & S. Abu Amr (Eds.), *Advanced Oxidation Processes (AOPs) in Water and Wastewater Treatment* (pp. 150–170). Hershey, PA: IGI Global. doi:10.4018/978-1-5225-5766-1.ch007

Mujere, N. (2016). Water Hyacinth: Characteristics, Problems, Control Options, and Beneficial Uses. In A. McKeown & G. Bugyi (Eds.), *Impact of Water Pollution on Human Health and Environmental Sustainability* (pp. 343–361). Hershey, PA: IGI Global. doi:10.4018/978-1-4666-9559-7.ch015

Mujere, N., & Isidro, M. (2016). Impacts of Artisanal and Small-Scale Gold Mining on Water Quality in Mozambique and Zimbabwe. In A. McKeown & G. Bugyi (Eds.), *Impact of Water Pollution on Human Health and Environmental Sustainability* (pp. 101–119). Hershey, PA: IGI Global. doi:10.4018/978-1-4666-9559-7.ch005

Nazir, R. (2017). Advanced Nanomaterials for Water Engineering and Treatment: Nano-Metal Oxides and Their Nanocomposites. In T. Saleh (Ed.), *Advanced Nanomaterials for Water Engineering, Treatment, and Hydraulics* (pp. 84–126). Hershey, PA: IGI Global. doi:10.4018/978-1-5225-2136-5.ch005

Obeng, P. A., Oduro-Kwarteng, S., Keraita, B., Bregnhøj, H., Abaidoo, R. C., Awuah, E., & Konradsen, F. (2016). On-Site Sanitation Systems for Low-Income Countries: Technical Guidelines for Groundwater Pollution and Nuisance Control. In A. McKeown & G. Bugyi (Eds.), *Impact of Water Pollution on Human Health and Environmental Sustainability* (pp. 319–342). Hershey, PA: IGI Global. doi:10.4018/978-1-4666-9559-7.ch014

Olajubu, E. A., Ajayi, G., Oke, I., & Asahiah, F. O. (2017). An ANN Model for Predicting the Quantity of Lead and Cadmium Ions in Industrial Wastewater. *International Journal of Information Communication Technologies and Human Development, 9*(4), 32–44. doi:10.4018/IJICTHD.2017100103

Paris, R. S. (2016). Issues with Water Quality: How Do We Get Our Fellow Citizens to Care? In A. McKeown & G. Bugyi (Eds.), *Impact of Water Pollution on Human Health and Environmental Sustainability* (pp. 199–214). Hershey, PA: IGI Global. doi:10.4018/978-1-4666-9559-7.ch009

Pedemonte, D. C., Frison, N., Tayà, C., Ponsa, S., & Fatone, F. (2017). Chemical and Biological Processes for Nutrients Removal and Recovery. In Á. Val del Río, J. Campos Gómez, & A. Mosquera Corral (Eds.), *Technologies for the Treatment and Recovery of Nutrients from Industrial Wastewater* (pp. 76–111). Hershey, PA: IGI Global. doi:10.4018/978-1-5225-1037-6.ch004

Prabu, S. L., Suriyaprakash, T., Kandasamy, R., & Rathinasabapathy, T. (2016). Effective Waste Water Treatment and its Management. In A. Rathoure & V. Dhatwalia (Eds.), *Toxicity and Waste Management Using Bioremediation* (pp. 312–334). Hershey, PA: IGI Global. doi:10.4018/978-1-4666-9734-8.ch016

Priyanka, M., Kumar, D., Shankar, U., Yadav, A., & Yadav, K. (2018). Agricultural Waste Management for Bioethanol Production. In V. Pathak, & Navneet (Eds.), Handbook of Research on Microbial Tools for Environmental Waste Management (pp. 1-33). Hershey, PA: IGI Global. doi:10.4018/978-1-5225-3540-9.ch001

Rajasekaran, R. N., Puddoo, H., & Nguyenhuynh, T. N. (2019). An Overview of Treatment of Antibiotics Using Advanced Oxidation Process. In H. Aziz & S. Abu Amr (Eds.), *Advanced Oxidation Processes (AOPs) in Water and Wastewater Treatment* (pp. 226–260). Hershey, PA: IGI Global. doi:10.4018/978-1-5225-5766-1.ch010

Rangreez, T. A., Mobin, R., Chisti, H., Bashir, R., & Ara, T. (2019). Advanced Water Treatment Systems and Their Applications. In A. Hussain & S. Ahmed (Eds.), *Advanced Treatment Techniques for Industrial Wastewater* (pp. 1–18). Hershey, PA: IGI Global. doi:10.4018/978-1-5225-5754-8.ch001

Rastogi, M., & Nandal, M. (2018). Composting an Emerging Technology for Solid Waste Management in India. In V. Pathak, & Navneet (Eds.), Handbook of Research on Microbial Tools for Environmental Waste Management (pp. 105-126). Hershey, PA: IGI Global. doi:10.4018/978-1-5225-3540-9.ch006

Rathoure, A. K. (2016). Heavy Metal Pollution and its Management: Bioremediation of Heavy Metal. In A. Rathoure & V. Dhatwalia (Eds.), *Toxicity and Waste Management Using Bioremediation* (pp. 27–50). Hershey, PA: IGI Global. doi:10.4018/978-1-4666-9734-8.ch002

Sadegh, H., & Ali, G. A. (2019). Potential Applications of Nanomaterials in Wastewater Treatment: Nanoadsorbents Performance. In A. Hussain & S. Ahmed (Eds.), *Advanced Treatment Techniques for Industrial Wastewater* (pp. 51–61). Hershey, PA: IGI Global. doi:10.4018/978-1-5225-5754-8.ch004

Said, S., Hussain, A., & Sharma, G. (2018). Water Quality Mapping of Yamuna River Stretch Passing Through Delhi State Using High Resolution Geoeye-2 Imagery. *International Journal of Applied Geospatial Research*, *9*(4), 23–35. doi:10.4018/IJAGR.2018100102

Saleh, T. A. (2017). An Overview of Nanomaterials for Water Technology. In T. Saleh (Ed.), *Advanced Nanomaterials for Water Engineering, Treatment, and Hydraulics* (pp. 1–12). Hershey, PA: IGI Global. doi:10.4018/978-1-5225-2136-5.ch001

Saleh, T. A., Adio, S. O., Parthasarathy, P., & Danmaliki, G. I. (2017). Scientific Insights Into Modified and Non-Modified Biomaterials for Sorption of Heavy Metals From Water. In T. Saleh (Ed.), *Advanced Nanomaterials for Water Engineering, Treatment, and Hydraulics* (pp. 13–39). Hershey, PA: IGI Global. doi:10.4018/978-1-5225-2136-5.ch002

Saleh, T. A., & Danmaliki, G. I. (2017). Polymer Consumption, Environmental Concerns, Possible Disposal Options, and Recycling for Water Treatment. In T. Saleh (Ed.), *Advanced Nanomaterials for Water Engineering, Treatment, and Hydraulics* (pp. 200–222). Hershey, PA: IGI Global. doi:10.4018/978-1-5225-2136-5.ch008

Saleh, T. A., Majeed, S., Nayak, A., & Bhushan, B. (2017). Principles and Advantages of Microwave-Assisted Methods for the Synthesis of Nanomaterials for Water Purification. In T. Saleh (Ed.), *Advanced Nanomaterials for Water Engineering, Treatment, and Hydraulics* (pp. 40–57). Hershey, PA: IGI Global. doi:10.4018/978-1-5225-2136-5.ch003

Satpathy, R. (2018). Microbial Bioreactor Systems for Dehalogenation of Organic Pollutants. In V. Pathak, & Navneet (Eds.), Handbook of Research on Microbial Tools for Environmental Waste Management (pp. 319-340). Hershey, PA: IGI Global. doi:10.4018/978-1-5225-3540-9.ch015

Shankara, S., Dupadahalli, K., Vijayakumar, M. H., & Gaddad, S. M. (2016). Decolorization of Direct Blue: 14 Dye by Thermoalkalophilic Aerobic Bacillus sp. In A. Rathoure & V. Dhatwalia (Eds.), *Toxicity and Waste Management Using Bioremediation* (pp. 279–294). Hershey, PA: IGI Global. doi:10.4018/978-1-4666-9734-8.ch014

Sharma, A., Kaur, M., Katnoria, J. K., & Nagpal, A. K. (2016). Heavy Metal Pollution: A Global Pollutant of Rising Concern. In A. Rathoure & V. Dhatwalia (Eds.), *Toxicity and Waste Management Using Bioremediation* (pp. 1–26). Hershey, PA: IGI Global. doi:10.4018/978-1-4666-9734-8.ch001

Shikuku, V. O., & Achieng', G. O. (2019). Occurrence and Fate of Selected Heavy Metals in a Conventional Municipal Wastewater Treatment Plant in Kisumu City, Kenya: A Case Study. In A. Hussain & S. Ahmed (Eds.), *Advanced Treatment Techniques for Industrial Wastewater* (pp. 211–224). Hershey, PA: IGI Global. doi:10.4018/978-1-5225-5754-8.ch012

Shikuku, V. O., Kowenje, C. O., & Nyairo, W. N. (2017). Fundamentals and Sources of Magnetic Nanocomposites and Their Sorption Properties. In T. Saleh (Ed.), *Advanced Nanomaterials for Water Engineering, Treatment, and Hydraulics* (pp. 58–83). Hershey, PA: IGI Global. doi:10.4018/978-1-5225-2136-5.ch004

Shikuku, V. O., Nyairo, W. N., & Kowenje, C. O. (2019). Preparation and Application of Biochars for Organic and Microbial Control in Wastewater Treatment Regimes. In A. Hussain & S. Ahmed (Eds.), *Advanced Treatment Techniques for Industrial Wastewater* (pp. 19–34). Hershey, PA: IGI Global. doi:10.4018/978-1-5225-5754-8.ch002

Shrivas, K., Ghosale, A., & Maji, P. (2017). Advanced Nanomaterials for the Removal of Chemical Substances and Microbes From Contaminated and Waste Water. In T. Saleh (Ed.), *Advanced Nanomaterials for Water Engineering, Treatment, and Hydraulics* (pp. 127–161). Hershey, PA: IGI Global. doi:10.4018/978-1-5225-2136-5.ch006

Simmonds, J., Gómez, J. A., & Ledezma, A. (2018). Statistical and Data Mining Techniques for Understanding Water Quality Profiles in a Mining-Affected River Basin. *International Journal of Agricultural and Environmental Information Systems*, *9*(2), 1–19. doi:10.4018/IJAEIS.2018040101

Singh, R., & Chauhan, M. (2016). Effective Management of Agro-Industrial Residues as Composting in Mushroom Industry and Utilization of Spent Mushroom Substrate for Bioremediation. In A. Rathoure & V. Dhatwalia (Eds.), *Toxicity and Waste Management Using Bioremediation* (pp. 158–177). Hershey, PA: IGI Global. doi:10.4018/978-1-4666-9734-8.ch008

Smith, M. C. (2016). Design of the Education Campaign: Creating a Motivated, Environmentally Concerned Public through Improved Scientific Literacy. In A. McKeown & G. Bugyi (Eds.), *Impact of Water Pollution on Human Health and Environmental Sustainability* (pp. 215–239). Hershey, PA: IGI Global. doi:10.4018/978-1-4666-9559-7.ch010

Starkl, M., Brunner, N., Hauser, A. W., Feil, M., & Kasan, H. (2018). Addressing Sustainability of Sanitation Systems: Can it be Standardized? *International Journal of Standardization Research*, *16*(1), 39–51. doi:10.4018/IJSR.2018010103

Stefanakis, A. I. (2016). Constructed Wetlands: Description and Benefits of an Eco-Tech Water Treatment System. In A. McKeown & G. Bugyi (Eds.), *Impact of Water Pollution on Human Health and Environmental Sustainability* (pp. 281–303). Hershey, PA: IGI Global. doi:10.4018/978-1-4666-9559-7.ch012

Stefanakis, A. I., & Becker, J. A. (2016). A Review of Emerging Contaminants in Water: Classification, Sources, and Potential Risks. In A. McKeown & G. Bugyi (Eds.), *Impact of Water Pollution on Human Health and Environmental Sustainability* (pp. 55–80). Hershey, PA: IGI Global. doi:10.4018/978-1-4666-9559-7.ch003

Stratigea, A., Leka, A., & Panagiotopoulou, M. (2017). In Search of Indicators for Assessing Smart and Sustainable Cities and Communities' Performance. *International Journal of E-Planning Research*, *6*(1), 43–73. doi:10.4018/IJEPR.2017010103

Tajarudin, H. A., Makhtar, M. M., Azmi, M. S., Zainuddin, N. I., & Ali, D. H. (2019). Introduction to Water and Wastewater Treatment. In H. Aziz & S. Abu Amr (Eds.), *Advanced Oxidation Processes (AOPs) in Water and Wastewater Treatment* (pp. 1–29). Hershey, PA: IGI Global. doi:10.4018/978-1-5225-5766-1.ch001

Tischler, D., Qi, J., Ngo, A. C., & Schlömann, M. (2018). Microbial Degradation of Azo Dyes: The Role of Azoreductase to Initiate Degradation. In V. Pathak, & Navneet (Eds.), Handbook of Research on Microbial Tools for Environmental Waste Management (pp. 341-371). Hershey, PA: IGI Global. doi:10.4018/978-1-5225-3540-9.ch016

Turcea, V. C., & Mihai, M. M. (2019). The Impact of CO2 Emissions From Waste on Human Health: A Consequence of Globalization. *International Journal of Sustainable Economies Management*, *8*(1), 17–28. doi:10.4018/IJSEM.2019010102

Umesh, M., & Basheer, T. (2018). Microbe Mediated Bioconversion of Fruit Waste Into Value Added Products: Microbes in Fruit Waste Management. In V. Pathak, & Navneet (Eds.), Handbook of Research on Microbial Tools for Environmental Waste Management (pp. 57-78). Hershey, PA: IGI Global. doi:10.4018/978-1-5225-3540-9.ch004

Val del Río, Á., Fernández, P. C., Gómez, J. L., & Mosquera-Corral, A. (2017). Nutrients Pollution in Water Bodies: Related Legislation in Europe and the United States. In Á. Val del Río, J. Campos Gómez, & A. Mosquera Corral (Eds.), *Technologies for the Treatment and Recovery of Nutrients from Industrial Wastewater* (pp. 21–42). Hershey, PA: IGI Global. doi:10.4018/978-1-5225-1037-6.ch002

Val del Río, Á., Fuentes, A. P., Giustinianovich, E. A., Gomez, J. L., & Mosquera-Corral, A. (2017). Anammox Process: Technologies and Application to Industrial Effluents. In Á. Val del Río, J. Campos Gómez, & A. Mosquera Corral (Eds.), *Technologies for the Treatment and Recovery of Nutrients from Industrial Wastewater* (pp. 264–289). Hershey, PA: IGI Global. doi:10.4018/978-1-5225-1037-6.ch010

Valverde-Pérez, B., Flores-Alsina, X., Vangsgaard, A. K., & Mauricio-Iglesias, M. (2017). Modelling and Control of Nitrogen and Phosphorus Removing Systems. In Á. Val del Río, J. Campos Gómez, & A. Mosquera Corral (Eds.), *Technologies for the Treatment and Recovery of Nutrients from Industrial Wastewater* (pp. 174–201). Hershey, PA: IGI Global. doi:10.4018/978-1-5225-1037-6.ch007

Verma, A., & Shukla, P. K. (2016). A Prospective Study on Emerging Role of Phytoremediation by Endophytic Microorganisms. In A. Rathoure & V. Dhatwalia (Eds.), *Toxicity and Waste Management Using Bioremediation* (pp. 236–265). Hershey, PA: IGI Global. doi:10.4018/978-1-4666-9734-8.ch012

Vipond, E., & Rahman, P. K. (2018). Anaerobic Digestion Enhancement With Microbial Electrolysis Cells: Is Biomethane Production the Direction to Go for Commercialization? In V. Pathak, & Navneet (Eds.), Handbook of Research on Microbial Tools for Environmental Waste Management (pp. 259-269). Hershey, PA: IGI Global. doi:10.4018/978-1-5225-3540-9.ch012

Voulvoulis, N., & Georges, K. (2016). Industrial and Agricultural Sources and Pathways of Aquatic Pollution. In A. McKeown & G. Bugyi (Eds.), *Impact of Water Pollution on Human Health and Environmental Sustainability* (pp. 29–54). Hershey, PA: IGI Global. doi:10.4018/978-1-4666-9559-7.ch002

Wahla, V., & Shukla, S. (2018). Role of Microorganisms in Bioremediation of Pesticides. In V. Pathak, & Navneet (Eds.), Handbook of Research on Microbial Tools for Environmental Waste Management (pp. 164-189). Hershey, PA: IGI Global. doi:10.4018/978-1-5225-3540-9.ch009

Willems, R. C., & MacDonald, S. A. (2016). Access to Safe Drinking Water: A Low Income Sustainable Water Purification System. In A. McKeown & G. Bugyi (Eds.), *Impact of Water Pollution on Human Health and Environmental Sustainability* (pp. 304–318). Hershey, PA: IGI Global. doi:10.4018/978-1-4666-9559-7.ch013

Zakaria, S. N. (2019). Ozonation With Catalyst in Landfill Leachate Treatment. In H. Aziz & S. Abu Amr (Eds.), *Advanced Oxidation Processes (AOPs) in Water and Wastewater Treatment* (pp. 324–354). Hershey, PA: IGI Global. doi:10.4018/978-1-5225-5766-1.ch012

Zaman, A., Das, P., & Banerjee, P. (2016). Biosorption of Dye Molecules. In A. Rathoure & V. Dhatwalia (Eds.), *Toxicity and Waste Management Using Bioremediation* (pp. 51–74). Hershey, PA: IGI Global. doi:10.4018/978-1-4666-9734-8.ch003

Ziemba, R., Cuker, B. E., Stein, J., Meuninck, R., & Wan, J. A. (2016). Health Professionals Can Protect Water Quality: Tools for Educators, Advocates, and Practitioners. In A. McKeown & G. Bugyi (Eds.), *Impact of Water Pollution on Human Health and Environmental Sustainability* (pp. 240–280). Hershey, PA: IGI Global. doi:10.4018/978-1-4666-9559-7.ch011

About the Authors

Athar Hussain is working as an Associate Professor in Department of Civil Engineering and Head, Civil and Environmental Engineering Department at Ch. B. P. Govt. College of Engineering, New Delhi. Dr. Hussain completed his Ph.D. in Environmental Engineering from IIT Roorkee in year 2007. Thereafter he joined as an Assistant Director with Environment Management Group at National Productivity Council, New Delhi till Oct. 2010. Thereafter he worked as an Assistant Professor, civil engineering department in School of Engineering at Gautam Buddha University till December 2015. His research area includes industrial wastewater treatment and solid waste management with special focus on anaerobic treatment of wastewater and solid waste. Dr. Hussain has more than 15 years of experience in the area of teaching, research and consultancy. He has also guided more than 60 M.Tech dissertations. He has published more than 90 papers in international referred journals and international conferences. He is also an Associate Member of The Institution of Engineers (IE) India. He has also handled various national and policy level projects with government and private organizations.

Index

P

S

T

W

www.ingramcontent.com/pod-product-compliance
Lightning Source LLC
LaVergne TN
LVHW080558300926
849162LV00005B/984
* 9 7 8 1 5 2 2 5 9 4 4 1 3 *